3 8006 10097 1764

CRANFIELD INSTITUTE OF TECHNOLOGY
LIBRARY

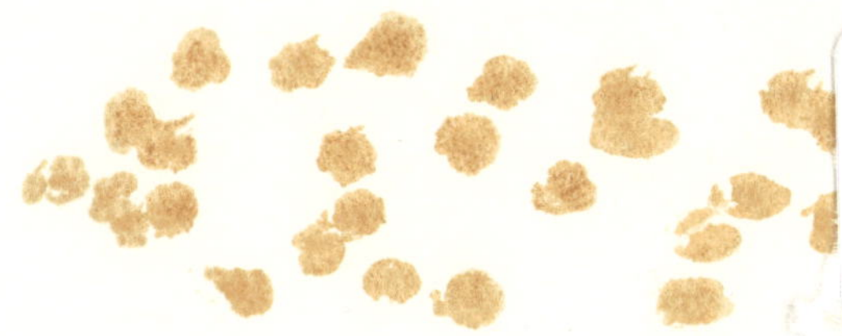

AF616260

PHOTONS AND QUANTUM FLUCTUATIONS

Other books in series

Chaos, Noise and Fractals
edited by E R Pike and L A Lugiato

Frontiers in Quantum Optics
edited by E R Pike and S Sarkar

Basic Methods of Tomography and Inverse Problems
edited by P C Sabatier

Nonlinear Phenomena and Chaos
edited by S Sarkar

MALVERN PHYSICS SERIES

General Series Editor: **Professor E R Pike** FRS

PHOTONS AND QUANTUM FLUCTUATIONS

Edited by

E R Pike
Royal Signals and Radar Establishment, Malvern and Department of Physics, King's College, London

and

H Walther
Sektion Physik, Universität München and Max-Planck-Institut für Quantenoptik, Garching

Adam Hilger, Bristol and Philadelphia

British Library Cataloguing in Publication Data

Photons and quantum fluctuations.
1. Quantum optics
I. Pike, E. R. (Edward Roy). II. Walther, H. (Herbert), *1935*– III. Series
535′.15

ISBN 0-85274-240-1

US Library of Congress Cataloging-in-Publication Data are available

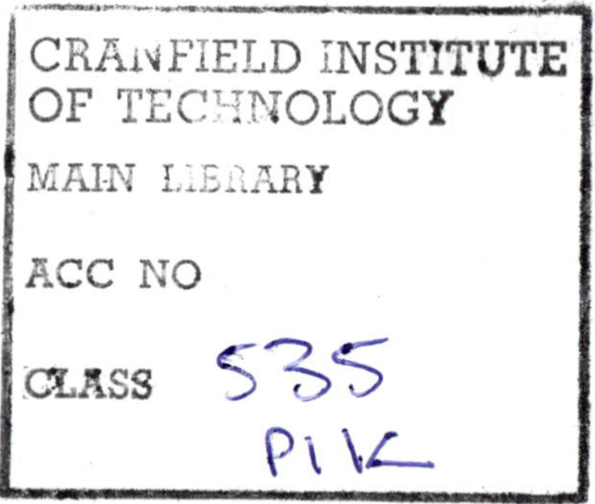

Published under the Adam Hilger imprint by IOP Publishing Ltd
Techno House, Redcliffe Way, Bristol BS1 6NX, England
242 Cherry Street, Philadelphia, PA 19106, USA

Printed in Great Britain by J W Arrowsmith Ltd, Bristol

CONTENTS

PREFACE

There has been considerable progress in the field of quantum optics in recent years. New experimental techniques have enlarged our basic understanding of radiation–matter coupling. Single particle events can now be studied in detail and it has become possible to explore non-classical properties of radiation not only in photon correlation measurements of antibunching (silent light!) but also via the dynamics of the radiation–atom coupling. In cavities electromagnetic fields characterized by fixed photon numbers (Fock states) can be generated and it is possible to study the interaction of atoms in such fields.

Furthermore, methods of nonlinear optics allow, via parametric down-conversion, two photons to be created simultaneously; by use of the photoelectric detection of one of them as a trigger, a good approximation to the ideal localized one-photon state can be achieved in the other. By means of nonlinear optical processes squeezed states can also be generated which allow measurements with precision beyond the limit set by the zero-point or vacuum fluctuations of the optical field. Among the potential applications of such fields is their use in a large laser interferometer for gravitational wave detection.

These exciting and recent developments together with other fundamental theoretical contributions in quantum optics are discussed in this book in detail. The different contributions review talks which were given at a special ONR Seminar held on January 21st and 22nd, 1988 at the Istituto d'Arte in Cortina d'Ampezzo, Italy, and at which we were honoured to join their 100th anniversary celebrations. The spectacular natural environment of Cortina was as inspiring and exciting as the topics of the seminar.

This Special Seminar was followed by a NATO ARW on 'Squeezed and Non-Classical Light', the proceedings of which will be published by Plenum Press. The two volumes may be regarded as complementary.

We would like to acknowledge, on behalf of all the participants, the generous support and encouragement of the US Office of Naval Research in London. We would also like to acknowledge additional financial assistance provided by Ministero Pubblica Istruzione, Università di Roma 'La Sapienza', Consiglio Nazionale delle Ricerche (CNR), Gruppo Nazionale di Struttura della Materia (CNR) and local sponsorship from Olivetti Spa, Consorzio per lo Sviluppo e Turismo di Cortina d'Ampezzo, Municipio di Cortina d'Ampezzo, Istituto Statale d'Arte and Hotel Europa. We express our gratitude for the invaluable assistance of Professor Paolo Tombesi of the University of Rome and the inspiration of Professor Danny Walls to suggest such a timely meeting. Special thanks are due to Professor M Spamponi, Mr G Milani, Ing E Cardazzi, Professor G Demenego, Mr E Demenego and, in particular, Professor G Olivieri. Jim Revill of Adam Hilger provided expert assistance with these proceedings and, last but by no means least, we thank our willing hard-worked secretaries Beverley James, Angela Di Silvestro and Marcella Mastrofini.

Roy Pike
Herbert Walther

May 1988

LIST OF CONTRIBUTORS

I ABRAM CNET
196 Avenue Henri Ravera
92200 Bagneux Cedex
France

M LE BERRE Laboratoire de PPM
Université de Paris Sud
Bâtiment 213
91405 Orsay Cedex
France

C FABRE Laboratoire de Spectroscopie Hertzienne de l'ENS
Université Pierre et Marie Curie
Tour 12, Premier Etage
4 Place de Jussieu
75252 Paris Cedex 05
France

E GIACOBINO Laboratoire de Spectroscopie Hertzienne de l'ENS
Université Pierre et Marie Curie
Tour 12, Premier Etage
4 Place de Jussieu
75252 Paris Cedex 05
France

F HAAKE Fachbereich Physik
Universität Essen – Gesamthochschule
Postfach 103764
4300 Essen 1
FRG

A HEIDMANN Laboratoire de Spectroscopie Hertzienne de l'ENS
Université Pierre et Marie Curie
Tour 12, Premier Etage
4 Place de Jussieu
75252 Paris Cedex 05
France

C K HONG

Department of Physics and Astronomy
University of Rochester
Rochester
New York
NY 14627
USA

R HOROWICZ

Laboratoire de Spectroscopie Hertzienne de l'ENS
Université Pierre et Marie Curie
Tour 12, Premier Etage
4 Place de Jussieu
75252 Paris Cedex 05
France

R LOUDON

Department of Physics
University of Essex
Wivenhoe Park
Colchester
CO4 3SQ
UK

L MANDEL

Department of Physics and Astronomy
University of Rochester
Rochester
New York
NY 14627
USA

G L MANDER

Department of Physics
University of Essex
Wivenhoe Park
Colchester
CO4 3SQ
UK

Z Y OU

Department of Physics and Astronomy
University of Rochester
Rochester
New York
NY 14627
USA

E R PIKE
Centre for Theoretical Studies
Royal Signals and Radar Establishment
St Andrews Road
Malvern
Worcestershire
WR14 3PS
UK

and
Department of Physics
King's College
Strand
London
WC2R 2LS
UK

J G RARITY
Royal Signals and Radar Establishment
St Andrews Road
Malvern
Worcestershire
WR14 3PS
UK

S REYNAUD
Laboratoire de Spectroscopie Hertzienne de l'ENS
Université Pierre et Marie Curie
Tour 12, Premier Etage
4 Place de Jussieu
75252 Paris Cedex 05
France

SARBEN SARKAR
Centre for Theoretical Studies
Royal Signals and Radar Establishment
St Andrews Road
Malvern
Worcestershire
WR14 3PS
UK

T J SHEPHERD
Royal Signals and Radar Establishment
St Andrews Road
Malvern
Worcestershire
WR14 3PS
UK

P R TAPSTER Royal Signals and Radar Establishment
St Andrews Road
Malvern
Worcestershire
WR14 3PS
UK

H WALTHER Sektion Physik
Universität München
Am Coulombwall 1
8046 Garching
FRG

and Max-Planck-Institut für Quantenoptik
8046 Garching
FRG

M WILKENS Fachbereich Physik
Universität Essen – Gesamthochschule
Postfach 103764
4300 Essen 1
FRG

H P YUEN Department of Electrical Engineering and Computer Science
Northwestern University
Evanston
Illinois 60620
USA

NONCLASSICAL LIGHT

H P YUEN

1. INTRODUCTION

A synopsis on nonclassical light will be presented, emphasizing the roles of different quantum amplifiers, the problem of overcoming the detrimental effect of loss, and highlighting certain points not explicitly brought out before. It is very far from a comprehensive review. In particular, no experiment will be described. A useful recent review on squeezed light with considerable scope has been given by Loudon and Knight (1987). No corresponding review on sub-Poissonian light is available, partly because it is as yet much less developed compared to squeezed light.

2. NONCLASSICAL LIGHT — SQUEEZING AND ANTIBUNCHING

The "vacuum" is filled with free electromagnetic field in its ground state. For a single mode with annihilation operator a and number operator $N \equiv a^\dagger a$, the ground state $|0>$ is an eigenstate of N as well as a. When a mean amplitude $\alpha \equiv \alpha_1 + i\alpha_2$ for the field quadratures a_1, a_2, $a \equiv a_1 + ia_2$, is added to the vacuum, we have a "classically" excited mode in a coherent state $|\alpha>$, $a|\alpha> = \alpha|\alpha>$ (Glauber 1963). A *classical state* is, by definition, a coherent state or a random superposition of coherent states, i.e., a state with a true probability density P-representation. Such states are obtained in all the conventional light sources. There are substantial quantum fluctuations even in a pure coherent state. The quadrature fluctuation in $|\alpha>$ is

$$\langle \Delta a_\phi^2 \rangle = \frac{1}{4}, \quad a_\phi \equiv a_1 \cos\phi - a_2 \sin\phi \tag{1}$$

while the photon number fluctuation is

$$\langle \Delta N^2 \rangle = \langle N \rangle = |\alpha|^2 \tag{2}$$

Further randomization as in the case of a general classical state could only increase these fluctuations.

In the phenomenon of *squeezing*, the quadrature fluctuation is reduced below the coherent state level (1),

$$\langle \Delta a_\phi^2 \rangle < \frac{1}{4} \tag{3}$$

for some ϕ. Any state exhibiting squeezing will be called a *squeezed state*; the two-photon coherent states (or squeezed states in the narrow sense) constitute the most important representative (Yuen 1976b). In (single-mode) antibunching, the number fluctuation is reduced below the Poisson level (2),

$$\langle \Delta N^2 \rangle < \langle N \rangle \tag{4}$$

Any state exhibiting strongly sub-Poissonian behavior, i.e., $\langle \Delta N^2 \rangle \ll \langle N \rangle$, will be called a *near-number state* (NNS); the number eigenstates $|n>$ constitute the ideal limiting case. Because of such reduced fluctuations, squeezed states and near-number states are not representable by true probability density P-representations. They are the paradigms of *nonclassical states*, i.e., states which are not classical. By definition, *nonclassical light* are light fields in nonclassical states. Thus far, the only nonclassical light of physical interest are squeezed state or near-number state light.

3. NONCLASSICAL STATES — TCS AND NUMBER STATES

A two-photon coherent state (TCS) $|\mu\nu\alpha>$ is an eigenstate of $\mu a + \nu a^\dagger$,

$$(\mu a + \nu a^\dagger)|\mu\nu\alpha>= (\mu\alpha + \nu\alpha^*)|\mu\nu\alpha>, \;\; |\mu|^2 - |\nu|^2 = 1 \tag{5}$$

The photocount statistics of $|\mu\nu\alpha>$ can be sub-Poissonian depending on the parameters. It is generally spiked at two photons apart; in particular, $\langle \mu\nu o|n\rangle \neq 0$ only for even n. The minimal quadrature fluctuation in $|\mu\nu\alpha>$ is

$$\langle \Delta a^2_{\phi_0} \rangle = \frac{1}{4}(|\mu| - |\nu|)^2 \tag{6}$$

with the corresponding maximal $\langle \Delta a^2_{\phi_0+\frac{\pi}{2}} \rangle = \frac{1}{4}(|\mu| + |\nu|)^2$ so that the uncertainty product

$$\langle \Delta a^2_{\phi} \rangle \langle \Delta a^2_{\phi+\frac{\pi}{2}} \rangle \geq \frac{1}{16} \tag{7}$$

achieves its minimum value at ϕ_0, which is a function of μ and ν (Yuen 1976b). The importance of TCS compared to other squeezed states in low noise application derives from the fact that the quadrature signal-to-noise ratio

$$\left(\frac{S}{N}\right)_{a_\phi} \equiv \frac{\langle a_\phi \rangle^2}{\langle \Delta a^2_\phi \rangle} \tag{8}$$

is maximized, under the fixed power constraint $tr\rho N \leq S$, by an appropriate TCS among all states (Yuen 1976a). In particular, the maximum value $(\frac{S}{N})^{TCS}_{a_\phi} = 4S(S+1)$ greatly exceeds the coherent state level $(\frac{S}{N})^{CS}_{a_\phi} = 4S$ for large S.

Since $\langle \Delta N^2 \rangle = 0$ in a number state $|n>$, it may appear that as

$$SNR_N \equiv \frac{\langle N \rangle^2}{\langle \Delta N^2 \rangle} \tag{9}$$

approaches infinity, near-number states are vastly superior to TCS. However, the discrete nature of $|n>$ greatly reduces this apparent advantage in most applications. In many ways, TCS and number states are comparable and complementary.

4. COHERENCE PROPERTIES

The characterization of the coherence properties of a field depends on the (quantum) measurement we are interested in performing on the field. In many applications, whether a field is coherent or not is irrelevant, particularly in regard to the so-called higher order coherence. (We do generally need a field traveling in a relatively well-defined direction.) Here, we limit ourselves to the usual first-order coherence describing interference, which is mathematically characterized by the factorization of the correlation function

$$G(r_1t_1; r_2t_2) \equiv tr\rho E^{\dagger}(r_1t_1)E(r_2t_2) \tag{10}$$

into the product of two c-number fields. Physically, this coherence condition expresses the spatial-temporal coherence of the field. It is equivalent (Titulaer and Glauber 1966) to having only one (general) space-time mode excited to an arbitrary state, while the other modes remain in vacuum. Thus, a number state field can be perfectly space-time coherent, as can a TCS field.

It is also well known that a multimode excited coherent state field is coherent to all orders (Glauber 1963). How does this reconcile with the above single mode condition? It turns out that a multimode coherent state field can be described, via a modal tranformation, as a single mode coherent state field. This possibility can be traced to the fact that the vacuum state is a coherent state. In a squeezed vacuum where all the modes are in $|\mu\nu 0>$ with the same μ, ν, a multimode excitation $|\mu\nu\alpha_i>$ would be equivalent to a single mode excitation in a similar fashion, while an ordinary multimode coherent state field would no longer be space-time coherent. This is just one illustration of the important fact that the vacuum state is a coherent state. On the other hand, although the vacuum state is also a number state, a multimode number state field is not equivalent to a single-mode one, i.e., not space-time coherent. This fact can be understood in terms of the unique linear transformation properties of Gaussian states, i.e., states with Gaussian characteristic functions. A detailed development will be given elsewhere.

5. GENERATION, PROPAGATION, AND DETECTION

The generation of TCS and NNS via nonlinear optical processes are, in some sense, surprisingly similar. This similarity is in addition to that between TCS and number-phase minimum uncertainty states when the nonclassical effects are small (Y. Yamamoto, N. Imoto, and S. Machida 1986). It seems to arise from an underlying intrinsic quantum correlation that is worth further exploring to great depth.

Nonclassical light do not propagate well. This is because significant loss, be it radiative, absorption, or otherwise, is often incurred during propagation and as we will see in the following section, ordinary linear loss re-introduces coherent state fluctuation into the field.

Each ordinary optical detection scheme corresponds to the quantum measurement of a certain observable — in direct detection one measures N, in homodyne detection a_ϕ, and in heterodyne detection a (Yuen and Shapiro 1980). Thus to utilize the advantage of a large (8) one homodynes, or balance-homodynes (Yuen and Chan 1983) for completely eliminating the local oscillator noise. To get a large (9) one counts photons. However, the effect of a nonunity quantum efficiency is equivalent to linear loss (Yuen and Shapiro 1978a). It seriously degrades the nonclassical effect as follows.

6. LOSS AND AMPLIFICATION

The effect of linear loss on a mode a can be represented by the transformation to another mode b (Yuen 1975, Yuen and Shapiro 1978b, Yuen 1983),

$$b = \eta^{\frac{1}{2}} a + (1-\eta)^{\frac{1}{2}} d \tag{11}$$

where the d-mode would be in vacuum except when a special medium is involved (Yuen 1975). It follows immediately from (11) that the resulting quadrature fluctuation is, for arbitrary a-mode state,

$$\langle \Delta b_\phi^2 \rangle = \eta \langle \Delta a_\phi^2 \rangle + (1-\eta)/4 \geq (1-\eta)/4 \tag{12}$$

Thus, a noise *floor* is introduced in addition to the usual signal attenuation, thereby greatly reducing the SNR advantage of a TCS a-mode. Similarly, the d-mode would introduce partition noise into the b-mode when the a-mode is in a number state. From (11) we have, for any a-mode state,

$$\langle \Delta N_b^2 \rangle = \eta^2 \langle \Delta N_a^2 \rangle + \eta(1-\eta)\langle N_a \rangle \tag{13}$$

The excess noise term $\eta(1-\eta)\langle N_a \rangle$ in (13) greatly reduces the noise advantage of NNS.

One general way to alleviate the effect of loss on squeezed states and NNS is to use appropriate optical preamplification. Post-amplification after loss could bring up the signal energy but could *not* recover the original SNR that has been degraded by the loss. There are three types of amplifiers corresponding to the generation of coherent state, TCS, and number state light. If, before taking loss, whether propagation or detection, one feeds an optical signal in any state into a proper amplifier matching the detection scheme, the detected signal-to-noise ratio could be made arbitrarily close to that of the amplifier input. To show this we need to examine the characteristics of these amplifiers.

7. PHASE INSENSITIVE LINEAR AMPLIFIERS

Let a and b be the input and output modal photon annihilation operators of an amplifier. For an ideal linear phase-insensitive amplifier (PIA), a and b are related by (Haus and Mullen 1962, Yuen 1983)

$$b = G^{\frac{1}{2}}a + (G-1)^{\frac{1}{2}}d^{\dagger} \tag{14}$$

where d is the annihilation operator for a vacuum mode. The added noise term in (14) that involves $d^{\dagger}$ arises directly from commutator preservation $[b, b^{\dagger}] = [a, a^{\dagger}] = I$. From (14) one immediately obtains

$$\langle \Delta b_{\phi}^2 \rangle = G\langle \Delta a_{\phi}^2 \rangle + (G-1)/4 \tag{15}$$

which shows that any squeezing in the input would disappear at the output for $G \geq 2$. A straightforward computation from (14) leads to

$$\langle \Delta N_b^2 \rangle - \langle N_b \rangle = G^2\langle \Delta N_a^2 \rangle + G(G-2)\langle N_a \rangle + (G-1)^2 \tag{16}$$

which shows that any sub-Poissonian statistics in a would disappear in b for $G \geq 2$ also. These features are, of course, confirmed in detailed calculations involving more explicit linear amplifier models (Hong, Friberg, and Mandel 1985; Mander, Loudon, and Shepherd, this volume).

It is clear that PIA would seriously degrade the SNR for TCS or NNS. From (14)-(15),

$$SNR_{b_{\phi}} = SNR_{a_{\phi}} \Big/ \left[1 + \frac{G-1}{G}\frac{1}{4\langle \Delta a_{\phi}^2 \rangle}\right] \tag{17}$$

Thus, the output SNR is greatly reduced from that of the input for strongly squeezed light. For coherent state input, $SNR_{b_{\phi}} \sim \frac{1}{2}SNR_{a_{\phi}}$. Similarly,

$$SNR_{N_b} = SNR_{N_a} \Big/ \left[1 + \frac{G-1}{G}\frac{\langle N_a \rangle + 1}{\langle \Delta N_a^2 \rangle}\right] \tag{18}$$

so that the output SNR is seriously degraded for strongly sub-Poissonian input. For coherent state input, $SNR_{N_b} \sim \frac{1}{2}SNR_{N_a}$. It can be readily shown that PIA would preserve the heterodyne signal-to-noise ratio

$$SNR_b = SNR_a \tag{19}$$

for arbitrary input state. In fact, linear amplification followed by classical measurement can be used in lieu of heterodyning, as the output quantum fluctuation is negligible compared to the output signal for large G. Since heterodyining (but *not* the TCS-hetrodyning of Yuen and Shapiro 1980) is not a useful measurement on TCS or NNS, we need other amplifiers to amplify nonclassical light in order to avoid serious degradation on SNR_{a_ϕ} or SNR_{N_a}.

8. PHASE SENSITIVE LINEAR AMPLIFIERS

An ideal phase sensitive linear amplifier (PSA) or parametric amplifier can be represented by (Yuen 1976b, Yuen 1983)

$$b_\phi = G^{\frac{1}{2}} a_\phi, \qquad b_{\phi+\frac{\pi}{2}} = G^{-\frac{1}{2}} a_{\phi+\frac{\pi}{2}} \tag{20}$$

where G is the gain for the quadrature a_ϕ. It follows immediately from (20) that a PSA preserves the homodyne detection SNR for any quadrature of the input in arbitrary state,

$$SNR_{b_\phi} = SNR_{a_\phi} \tag{21}$$

In spite of (21), it may be observed that squeezing would be destroyed at the output whenever $G \geq 1/4\langle \Delta a_\phi^2 \rangle$ as a result of (20). In most applications, however, it is the SNR and not the absolute fluctuation that matters. Thus, the useful advantage of squeezing could be preserved after phase-sensitive amplification. In fact, it would be *easier* to deal with a loss-insensitive and noise-insensitive amplified quadrature than a phase-sensitive squeezed quadrature.

To see how appropriate preamplification can overcome subsequent loss on an input a, let the PSA on output b pass through a linear attenuator of the form (11) with final output c. It is easily computed that (Yuen 1987)

$$SNR_{c_\phi} = SNR_{a_\phi} / \left[1 + \frac{\eta}{G(1-\eta)} \frac{1}{4\langle \Delta a_\phi^2 \rangle} \right] \tag{22}$$

where η can represent any loss, say a propagation loss or a detector quantum efficiency. From (22),

$$SNR_{c_\phi} \sim SNR_{a_\phi} \tag{23}$$

whenever G is sufficiently large. Note that the output SNR of an amplifier cannot be larger than that of the input, so that (23) represents the behavior of a "noiseless amplifier". Detection device noise can also be overcome by preamplification in a similar manner. If a PIA is used in place of the PSA, we would have $SNR_{c_1} \ll SNR_{a_1}$ for strongly squeezed input and $SNR_{c_1} \sim \frac{1}{2} SNR_{a_1}$ for coherent state input.

9. PHOTON NUMBER AMPLIFIERS

A noiseless photon amplifier or photon number amplifier (PNA) is a device that ideally would transform an input number state $|n>$ to an output number state $|Gn>$ for an integer G. At present, PNA is merely a device concept (Yuen 1986a, Yuen 1986b). No experimental result has been reported although a few schemes have been suggested. For an arbitrary input state, it is easily computed that the direct detection SNR is preserved in a PNA,

$$SNR_{N_b} = SNR_{N_a} \tag{24}$$

On the other hand, the input Fano factor is amplified by G,

$$F_b \equiv \frac{\langle \Delta N_b^2 \rangle}{\langle N_b \rangle} = GF_a \tag{25}$$

so that the output mode is no longer sub-Poissonian when $F_a > 1/G$.

Similar to the PSA case, let the PNA output b pass through a linear attenuator of final output c, with the resulting (Yuen 1986b)

$$SNR_{N_c} = SNR_{N_a} / \left[1 + \frac{\eta}{G(1-\eta)} \frac{\langle N_a \rangle}{\langle \Delta N_a^2 \rangle} \right] \tag{26}$$

From (26),

$$SNR_{N_c} \sim SNR_{N_a} \tag{27}$$

for a sufficiently large G. If a PIA is used in place of the PNA, we would have $SNR_{N_c} \ll SNR_{N_a}$ for NNS input and $SNR_{N_c} \sim \frac{1}{2} SNR_{N_a}$ for coherent state input. Again, detection device noise can be suppressed in a similar manner.

Equations (19), (21) and (24) show that *different amplifiers are suited for different detection schemes, regardless of the nature of the source.* Of course, different detection schemes are naturally suited for different nonclassical sources. We summarize this situation in the following table:

SOURCE	DETECTION	AMPLIFIER
Coherent state	heterodyne	PIA
TCS	homodyne	PSA
NNS	direct	PNA

It should be clear that PNA is indeed a natural amplifier, completing the triad for matching the three standard detection schemes. *We need PSA to preserve the advantage of squeezing, and PNA that of antibunching.* In addition, PSA and PNA are useful even for coherent state sources, as they are still 3db superior to PIA in homodyne or direct detection that are sometimes usefully employed for detecting coherent state light. It should also be emphasized that they are necessary if one wants to overcome loss or detection noise by preamplification.

10. APPLICATIONS

Nonclassical light with matching detection leads to significant improvement in SNR, as we have seen. Even in situations where SNR is not the most appropriate performance measure, it is clear that the low noise characteristics of TCS and NNS, or simply their difference from classical states, could be useful. The many applications that have been suggested for them, to my knowledge, are listed in the following:

- Communications (Yuen 1975; Yuen and Shapiro 1978b, 1980; Shapiro, Yuen and Machado-Mata 1979; Yuen 1987).
- Fiber Tapping (Shapiro 1980; Yuen 1987).
- Gyros (Dorschner, Haus, et. al. 1980).
- Interferometry (Caves 1981; Bondurant and Shapiro 1985; Yuen 1986b; Yurke, McCall and Klauder 1986).
- Optical Computing (Yamamoto).
- Optical Memory (Levenson).
- Precision Measurements (Yuen 1976b; Kimble; Slusher).
- Spectroscopy (Gardiner 1986; Milburn 1986; Yurke and Whittaker 1987).

REFERENCES

Bondurant, R. S. and Shaprio, J. H. 1984, Phys. Rev. D, 30, 2548.

Caves, C. M., 1981, Phys. Rev. D, 23, 1693.

Dorschner, T. A. Haus, H. A., et al, 1980, IEEE J. Quant. Electron., 16, 1376.

Gardiner, C.W. 1986, Phys. Rev. Lett., 56, 1917.

Glauber, R. J., 1963, Phys. Rev., 131, 2766.

Haus, H.A., and Mullen, J. A., 1963, Phys. Rev., 128, 2407.

Hong, C. K., Friberg, S., and Mandel, L., 1985, JOSA-B, 2, 494.

Loudon, R., and Knight, P. L., 1987, J. Mod. Opt., 34, 709.

Mander, G., Loudon, R., and Shepherd, T., this volume.

Milburn G. J., 1986, Phys. Rev. A, 34, 4882.

Shapiro, J. H., 1980, Optics Lett., 5, 351.

Shapiro, J. H., Yuen, H. P., and Machado-Mata, J. A., 1979, IEEE Trans. Inform. Theory, 25, 179.

Titulaer, U. M., and Glauber, R. J., 1966, Phys. Rev., 145, 1041.

Yamamoto, Y., Imoto, N., and Machida, S., 1986, Phys. Rev. A, 33, 3243.

Yuen, H. P., 1975, Proceedings of the 1975 Conference on Information Sciences and Systems, John Hopkins University Press, Baltimore, pp. 171-177.

Yuen, H. P., 1976a, Phys. Lett. A, 56, 101.

Yuen, H. P., 1976b, Phys. Rev. A, 13, 2226.

Yuen, H. P., 1983, Quantum Optics, Experimental Gravitation, and Measurement Theory, P. Meystre and M. O. Scully, Eds., Plenum, New York, pp. 249-268.

Yuen, H. P., 1986a, Phys. Lett. A, 113, 405.

Yuen, H. P., 1986b, Phys. Rev. Lett., 56, 2176.

Yuen, H. P., 1987, Optics Lett., 12, 789.

Yuen, H. P., and Chan, V. W. S., 1983, Optics Lett., 8, 177.

Yuen, H. P., and Shapiro, J. H., 1978a, Proc. of the Fourth Rochester Conf. on Coherence and Quantum Optics, L. Mandel and E. Wolf, Eds., Plenum, New York, pp. 719-727.

Yuen, H. P., and Shapiro, J. H., 1978b, IEEE Trans. Inform. Theory, 24, 657.

Yuen, H. P., and Shapiro, J. H., 1980, IEEE Trans. Inform. Theory, 26, 78.

Yurke, B., McCall, S. L., and Klauder, J. R., 1986, Phys. Rev. A, 33, 4033.

Yurke, B., and Whittaker, E. A., 1987, Optics Lett., 12, 236.

SINGLE-ATOM OSCILLATORS

H WALTHER

Modern methods of laser spectroscopy allow the study of single atoms or ions in an unperturbed environment. This has opened up interesting new experiments, among them the detailed study of radiation-atom coupling. In the following two experiments of this type are reviewed: the single-atom maser and the study of the resonance fluorescence of a single stored ion.

The simplest and most fundamental system for studying radiation-matter coupling is a single two-level atom interacting with a single mode of an electromagnetic field in a cavity. This problem received a great deal of attention shortly after the maser was invented. At that time, however, the problem was of purely academic interest: the matrix elements describing the radiation-atom interaction are usually too small, so that the field of a single photon is not sufficient to lead to an atom-field evolution time shorter than the other characteristic times of the system, such as the excited state lifetime, the time of flight of the atom through the cavity and the cavity mode damping time. It was therefore not possible to test experimentally the fundamental theories of radiation-matter interaction. These theories predict, however, some interesting and basic effects. These include the (a) modification of the spontaneous emission rate of a single atom in a resonant cavity, (b) the oscillatory energy exchange between a single atom and the cavity mode, and (c) the disappearance and quantum revival of optical nutation induced in a single atom by a resonant field.

The situation concerning the experimental testing of these basic effects has drastically changed in the last few years since frequency-tunable lasers now allow population of highly excited atomic states characterized by a high main quantum number n of the valence electron. These states are generally called Rydberg states since their energy levels can be described by the simple Rydberg formula. The highly excited atoms are very suitable for observing the quantum effects in radiation-atom coupling for three reasons. Firstly, these states are very strongly coupled to the radiation field (the induced transition rates between neighbouring levels scale as n^4). Secondly, these transitions are in the millimetre wave region, which allows low-order mode cavities that are still sufficiently large to ensure rather long interaction times. Finally, Rydberg states have relatively long lifetimes with respect to spontaneous decay. For reviews see Haroche and Raimond (1985) and Gallas, Leuchs, Walther and Figger (1985).

The strong coupling of Rydberg atoms to resonant radiation between neighbouring levels can be understood in terms of the correspondence principle: with increasing n the classical evolution frequency of the highly excited electron becomes identical with the transition frequency to the neighbouring level; the atom therefore corresponds to a large dipole oscillating with the resonance frequency. (The dipole moment is very large since the atomic radius scales as n^2.)

In order to understand the modification of the spontaneous emission rate in an external cavity, we have to remember that in quantum electrodynamics this rate is determined by the density of modes of the electromagnetic field at the atomic transition frequency ω_0. The vacuum density of modes per unit volume depends on the square of the frequency. If the atom is not in free space, but in a resonant cavity instead, the continuum of modes is changed into a spectrum of discrete modes with one of them being in resonance with the atom. Since there is energy dissipation within the cavity, a photon radiated at a well-defined frequency will be smeared out over the full spectral width $\Delta\omega_c$ of the resonant mode. The full width at half maximum $\Delta\omega_c$ is related to the cavity quality factor $Q = \omega_c/\Delta\omega_c$.

The spontaneous decay rate of the atom in the cavity γ_c is enhanced in relation to that in free space γ_f by a factor given by the ratio of the corresponding mode densities (V_c is the volume of the cavity):

$$\gamma_c/\gamma_f = \rho_c(\omega_0)/\rho_f(\omega_0) = 2\pi Q/V_c\omega_0^3 = Q\lambda_0^3/4\pi^2 V_c.$$

For low-order cavities in the microwave region one has $V_c \approx \lambda_0^3$; the spontaneous emission rate is thus roughly increased by a factor of Q in a resonant cavity; conversely, the decay rate decreases when the cavity is mistuned. In this case the atom cannot emit a photon, since the cavity is not able to accept it, and therefore the energy has to stay with the atom.

Recently, quite a few experiments have been conducted with Rydberg atoms to demonstrate the enhancement and inhibition of spontaneous decay in external cavities or cavity-like structures. For the most recent experiment see Ihe et al. (1987).

There are also more subtle effects due to the change of the mode density: radiation corrections such as the Lamb shift and the anomalous magnetic dipole moment of the electron are also modified with respect to the free space value if they are calculated under the boundary conditions of a cavity (Barton 1987). The change is just of the order of magnitude of present experimental accuracy. Roughly speaking, one can say that these effects are determined by virtual transitions and not by real transitions as in the case of spontaneous decay.

In the following, attention is focused on discussing the one-atom maser in which the idealized case of a two-level atom interacting with a single mode of a radiation field is realized; the theory of this system was treated by Jaynes and Cummings (1963) many years ago. We concentrate on the dynamics of the atom-field interaction predicted by this theory. Some of the features are explicitly a consequence of the quantum nature of the electromagnetic field: the statistical and discrete nature of the photon field leads to new dynamic characteristics such as collapse and revivals in the Rabi nutation.

First we review the main results of the Jaynes-Cummings model with respect to the atomic dynamics. We consider a two-level atom in the excited state which enters a resonant cavity with a field of n photons. The probability $P_{e,n}$ of the atom to be in the excited state is then given by

$$P_{e,n}(t) = 1/2\ \{1+\cos[2\Omega(n+1)^{1/2}t]\}$$

where Ω is the single-photon Rabi frequency. With a fluctuating number of photons initially present in the cavity, the quantum Rabi solutions needs to be averaged over the probability distribution p(n) of having n photons in the mode at t = 0:

$$P_e(t) = 1/2 \sum_{n=0}^{\infty} p(n)\ \{1+\cos[2\Omega(n+1)^{1/2}t]\}$$

At a low atomic-beam flux, the cavity contains essentially thermal photons and their number is a random quantity conforming to Bose-Einstein statistics. In this case p(n) is given by $P_{th}(n) = \bar{n}^n{}_{th}/(\bar{n}_{th} + 1)^{n+1}$, with the average number of thermal photons being $\bar{n}_{th} = [\exp(h\nu/kT) - 1]^{-1}$. The distribution of Rabi frequencies results in an apparent random oscillation $P_{e,th}$. At higher atomic-beam fluxes the number of photons stored in the cavity increases and their statistics changes. If a coherent field is prepared in the cavity at t = 0, the probability distribution p(n) is given by a Poissonian: $p_c(n) = \exp(-\bar{n})\bar{n}^n/n!$ The Poisson spread in n gives a dephasing of the Rabi oscillations, and therefore $P_{e,c}(t)$ first exhibits a collapse. This is described in the resonant case by the approximate envelope $\exp(-\Omega^2 t^2/2)$ and is independent of the average photon number (this independence does not hold for nonresonant excitation). The collapse was also noted later in other work. After the collapse there is a range of interaction times for which $P_{e,c}(t)$ is independent of time. Later $P_{e,c}(t)$ then exhibits recorrelations (revivals) and starts oscillating again in a very complex way. As has been shown by Eberly and co-workers the recurrences occur at times given by $t = kT_R$ $(k = 1,2,...)$, with $T_R = 2\pi(\bar{n})^{1/2}/\Omega$ (Eberly et al., 1980, Narozhny et al., 1981, Yoo et al., 1981, Yoo et al., 1983). Both collapse and revivals in the coherent state are purely quantum features and have no classical counterpart.

The inversion also collapses and revives in the case of a chaotic Bose-Einstein field (Knight and Radmore, 1982). Here the photon-number spread is far larger than for the coherent state and the collapse time is much shorter. In addition, the revivals completely overlap and interfere to produce a very irregular time evolution. A classical thermal field represented by an exponential distribution of the intensity also shows collapse, but no revivals. Therefore the revivals can be considered as a clear quantum feature, but the collapse is less clear-cut as a quantum effect.

It is interesting to mention that in the case of two-photon processes the Rabi frequency turns out to be $2\Omega(n+1)$ rather than $2\Omega(n+1)^{1/2}$, enabling the sums over the photon numbers in $P_e(t)$ to be carried out in simple closed form. In this case the inversion revives perfectly with a completely periodic sequence (Knight, 1986).

The experimental setup used for the experimental test of the Jaynes-Cummings model is shown on Fig. 1. The atoms are injected into a superconducting cavity in the upper state of the maser transition. This was the 63 $p_{3/2}$ Rydberg level, populated by excitation with frequency-doubled light of a dye laser. The atoms are monitored by using field ionization. This detection can be performed state-selectively by choosing the proper field strength.

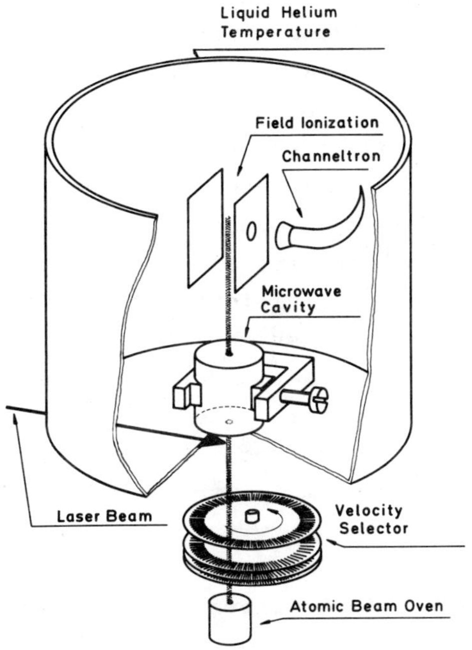

Fig. 1 Scheme of the single-atom maser for measuring quantum collapse and revival (Rempe, Walther, Klein, 1987)

In most of the experiments (Meschede, Walther and Müller, 1985; Rempe, Walther and Klein, 1987) the transition 63 $p_{3/2}$ - 61 $d_{5/2}$ with a frequency of 21.456 GHz was investigated. When the cavity is tuned in resonance to this transition, the number of atoms in the upper state decreases owing to enhanced spontaneous emission. The tuning of the cavity is performed by squeezing the cavity with piezoelectric elements. The flux of atoms is very low so that the average number of atoms in the cavity at a time is usually less than unity. The interaction time of the atoms with the cavity field can be varied by means of a Fizeau velocity selector. In this way, the dynamics of the energy exchange between the atom and cavity field can be investigated.

With very low atomic-beam flux, the cavity contains essentially thermal photons only. Their number is a random quantity according to Bose-Einstein statistics. When the velocity of the atoms is changed, the probability of the atom being in the excited state $P_e(t)$ after interaction varies with the interaction time in an apparently random way. At higher atomic-beam fluxes the atoms deposit energy in the cavity and the maser reaches the threshold so that the number of photons stored in the cavity increases and their statistics changes. For the case of a coherent field the probability distribution is a Poissonian. This distribution spread in n results in dephasing of the Rabi oscillations, and therefore the envelope of $P_e(t)$ collapses; after the collapse $P_e(t)$ starts oscillating again in a very complex way. As mentioned above, both collapse and revivals in the coherent state are pure quantum features.

The above-mentioned effects have been demonstrated experimentally. The experimental results clearly show the collapse and revival predicted by the Jaynes-Cummings model. Figure 2 shows a series of measurements obtained with the single-atom maser (Rempe, Walther and Klein, 1987). Plotted is the probability $P_e(t)$ of finding the atom in the upper maser level for increasing atomic flux N. The strong variation of $P_e(t)$ for interaction times between 50 and 80 s disappears for larger N and a revival shows up for N = 3000 s^{-1} for interaction times larger than 140 s. The average photon number in the cavity varies between 2.5 and 5, about 2 photons being due to the black-body field in the cavity corresponding to a temperature of 2.5 K.

There is another aspect of the single-atom maser which is very interesting: the non-classical statistics of the photons in the cavity. This problem is briefly discussed in the following. There are two approaches to the quantum theory of the one-atom maser. Filipowicz et al. (1986) use a microscopic approach to describe the device. On the other hand, Lugiato et

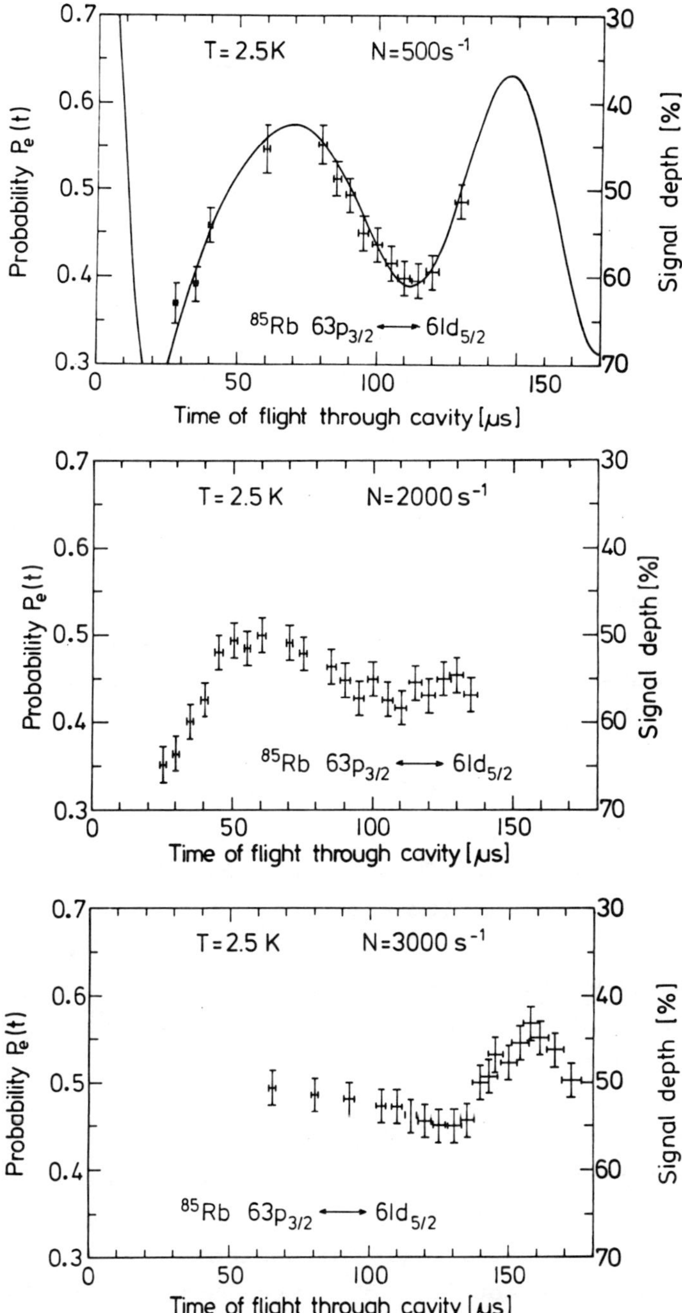

Fig. 2 Quantum collapse and revival in the one-atom maser. Plotted is the probability $P_e(t)$ of finding the atom in the upper maser level for different fluxes N of the atomic beam.

al. (1987) show that the standard macroscopic quantum laser theory leads to the same steady-state photon number distribution. The special features of the micromaser were not emphasized in the standard laser theory because the broadening due to spontaneous decay obscured the Rabi cycling of the atoms. When similar averages in the microscopic theory associated with inhomogeneous broadening are performed, equivalent results are obtained.

Both theoretical approaches predict that the statistics of the photons in the maser cavity depend on the interaction time of the atoms in the cavity. It turns out that the distribution is mostly sub-Poissonian. In the case of a high Q value on the order of $Q \approx 5 \times 10^{10}$ a state of the radiation field with a fixed photon number, i.e. a Fock state is expected (Krause, Scully and Walther, 1987).

To achieve such a state experimentally, two conditions have to be met. The first concerns the temperature; thermal photons have to be suppressed because they not only induce statistical decay but also result in superposition of number states. We can eliminate thermal photons by cooling the cavity to a low enough temperature. The second condition is that photons stored in the cavity for the duration of the experiment should not be lost, i.e. one needs a cavity in which losses can be neglected for this time, which means that the Q value has to be higher than 10^{10}. Both conditions can now be realized in the Rydberg maser experiment.

The results obtained for the single-atom maser also give a new insight into the statistical properties of masers and lasers. It was demonstrated that the photon statistics in the one-atom maser is in general sub-Poissonian or corresponds to a Fock state, in contrast to the usual masers and lasers, where coherent radiation, i.e. Poissonian statistics, is observed. The reason for this is that there is much stronger damping present in the cavities of the usual masers and lasers than in the micromaser; in addition, the atomic or molecular transitions also usually show stronger damping than the Rydberg states, this being especially the case for laser systems. Furthermore, the selected velocity of the atoms used in connection with the collapse and revival measurements (Rempe, Walther and Klein, 1987) in the micromaser leads to fixed interaction times in the cavity; this also helps to reduce the photon number fluctuations since the photon exchange between the atom and cavity field can be exactly controlled. The smallest fluctuations are achieved when the atoms leave the cavity again in the upper state. Of course, it is necessary that energy be deposited in the cavity in order to maintain maser oscillation, but the losses are very small with a high-Q-cavity and P_e for atoms leaving the cavity can be chosen very close to unity.

There are other interesting aspects of Rydberg masers which can only be briefly mentioned here. Recently, a two-photon maser was realized by Brune et al. (1987). The two-photon transition was chosen such that there is an intermediate level of suitable parity nearly halfway between the upper and lower maser levels, thus enhancing the transition amplitude. In such a device new features not present in one-photon masers can be observed, e.g. delayed start-up time at threshold or multistable behaviour. Unlike the one-photon maser, which behaves at threshold similarly to a 2nd-order phase transition, the two-photon maser is analogous to a 1st-order phase transition. At this point we should also mention that it was pointed out by Meystre and co-workers that the micromaser can be used to investigate aspects of chaos and problems of measurement theory (Meystre, 1987).

Another problem in radiation-matter interaction which has received a lot of interest is resonance fluorescence. Until the advent of the laser, light sources for spectroscopy consisted of ordinary spectral lamps excited by DC or RF discharges which produced light having a very broad spectral width, and hence very short correlation time, and a relatively low intensity. For such fields both the experimental and theoretical results are in general well understood. However, the development of the laser provided light sources which are sufficiently intense to saturate an atomic (or molecular) transition very easily. In addition, lasers are highly monochromatic, having a coherence time much greater than typical natural lifetimes of excited atomic states, and, latterly, tunable, allowing selective excitation of particular atomic transitions. As might be expected, it has been found that many interesting new phenomena are associated with such fields interacting with atomic systems.

The theoretical analysis of this new physical situation requires the use of techniques more general than those found adequate in the case of thermal fields. In the latter case the weakness of the atom-field interaction meant that perturbative techniques were generally sufficient. These techniques were based on the assumption that the initial state of the atomic system is essentially unchanged by the interaction. However, as saturation can easily be achieved with an intense laser field, more general non-perturbative methods are required. Furthermore, for a highly coherent field, one cannot consider successive photon emissions and absorption processes as being independent, since it is now possible for an atomic system to undergo many such processes during the correlation time of the laser field, and hence phase memory effects cannot be neglected.

Although a wide range of problems, both theoretical and experimental, involving laser fields have been studied, attention is confined here to just one aspect: the interaction of intense monochromatic light with an atomic system in which it is the properties of the fluorescent light which are of principal interest. Unlike in the problem discussed in connection with the one-atom maser, here the atom decays under the influence of the vacuum fluctuations of free space and the emitted photons disappear and cannot be reabsorbed.

Most of the ongoing work is concerned with the problem of

theoretically and experimentally determining the spectrum of the fluorescent light radiated by a two-level atom driven by an intense monochromatic field. This is the situation that gives rise to a dynamic Stark effect in which, for sufficiently strong fields, it is found that the spectrum of the scattered light splits into three peaks consisting of a central peak, centred at the driving field frequency with a width $\gamma/2$ ($1/\gamma$ being the Einstein A coefficient) and having a height three times that of two symmetrically placed sidebands, each of width $3\gamma/4$ and displaced from the central peak by the Rabi frequency. In addition, there is a delta-function (coherent) contribution, also positioned at the driving frequency. In the limit of strong driving fields, the energy carried by this last contribution is negligible in relation to the three-peak contribution. This result was first predicted by Mollow and was later experimentally confirmed very well (for reviews see Cohen-Tannoudji (1977) and Cresser, Häger, Leuchs, Rateike, Walther (1982)).

However, it is not only the spectral property of the fluorescent light that has come under investigation. Examination of the intensity correlation of the scattered field in the basic two-level atom has also attracted much attention since fluorescent light exhibits interesting statistical properties, especially when there is only a single atom at a time interacting with the laser beam; under those conditions the phenomenon of photon-antibunching can be observed. The single-atom condition cannot easily be fulfilled if the atoms of an atomic beam are observed. However, the new techniques of laser spectroscopy of single ions in a radio-frequency trap are very suitable for this purpose. This was recently demonstrated by Diedrich and Walther (1987). These experiments will be discussed in the following.

To study photon statistics usually the normalized intensity correlation function $g^{(2)}(\tau) = \langle I(t)\ I(t+\tau)\rangle / \langle I(t)\rangle^2$ is measured; this is proportional to the probability of a second photon being detected after a first one within a time interval

τ. For thermal and noncoherent light this probability has a maximum for $\tau = 0$ and decreases for larger τ. This behaviour is called photon bunching since it indicates that the photons occur in "clusters". Coherent light displays a value for the intensity correlation that is independent of τ. Quantized fields show additional dependences: the smallest value can occur at $\tau = 0$; this is termed antibunching. Such a field is produced by, for example, a single stored ion in the following way: after a photon is emitted the trapped ion returns to the ground state; before the next photon can be emitted, the ion has to be excited again. This happens through Rabi nutation in the external laser field. On the average a time of half a Rabi period has to elapse until another photon can be observed. The probability of two photons being emitted a short time after each other is therefore very small (Carmichael and Walls, 1976).

Previous experiments to investigate antibunching in resonance fluorescence have been performed by means of laser-excited collimated atomic beams. The initial results obtained by Kimble, Dagenais and Mandel (1977) showed for the second-order correlation function $g^{(2)}(\tau)$ a positive slope characteristic of photon antibunching, but $g^2(0)$ was larger than $g^2(\tau)$ for $\tau \rightarrow \infty$. This was due to number fluctuations in the atomic beam and to the finite interaction time of the atoms (Jakeman, Pike, Pusey and Vaugham, 1977; Kimble, Dagenais and Mandel, 1978). Later the analysis of the experiment was refined by Dagenais and Mandel (1978). Another experiment with longer interaction time was performed by Rateike, Leuchs and Walther (see Cresser et al., 1982). In the latter experiment the photon correlation was also measured for very low laser intensities.

The fluorescence of a single ion should also display the following property: the probability distribution of the photon number recorded in a finite time interval δt is narrower than Poissonian, which means in other words that the variance is smaller than the mean value of the photon number. This is because the single ion can only emit a single photon. Antibunching and sub-Poissonian statistics are often associated. They are, nevertheless, distinct properties and need not necessarily be simultaneously observed (Short and Mandel, 1983) as is the case in the experiment described here. Although there is evidence of antibunching in the atomic-beam experiments, the photon counts were not sub-Poissonian as a result of fluctuations in the number of atoms. In the experiment by Short and Mandel (1983) this effect was excluded by use of a special trigger scheme for the single-atom event. In the single ion setup these precautions are not necessary since there are no fluctuations in the atomic number.

The Paul trap used in the experiment is shown in Fig. 3. The trap is mounted inside a stainless-steel ultrahigh-vacuum chamber. Its ring diameter of 5 mm and its pole cap separation of 3.54 mm are much larger than for other single-ion radiofrequency traps. The single ion is closely confined in the centre of the trap by photon recoil cooling. For this purpose the laser frequency is tuned to the lower frequency side of the atomic resonance line. As a consequence, the trapped ion is only excited when it moves towards the laser beam the kinetic energy being simultaneously diminished by the linear momentum of the absorbed photons. The large size of the trap affords a large solid angle for detecting the fluorescence radiation. This radiation is transmitted through a molybdenum mesh covering a conical bore in the upper pole cap. The experiment was performed with $^{24}Mg^{+}$ ions. They were produced by means of an atomic beam which was ionized in the centre of the trap by an electron beam. The resonance transition $3^2S_{1/2} \rightarrow 3^2P_{3/2}$ has the wavelength $\lambda = 280$ nm.

To investigate the photon statistics, the second-order correlation function (intensity correlation) was measured in a Hanbury-Brown and Twiss-Experiment. The results for the intensity correlation of a single stored ion are shown in Fig. 4. Plotted is $g_I^{(2)}(\tau)$. Owing to a time delay in one of the signal channels the intensity correlation $g_I^{(2)}(\tau)$ could also

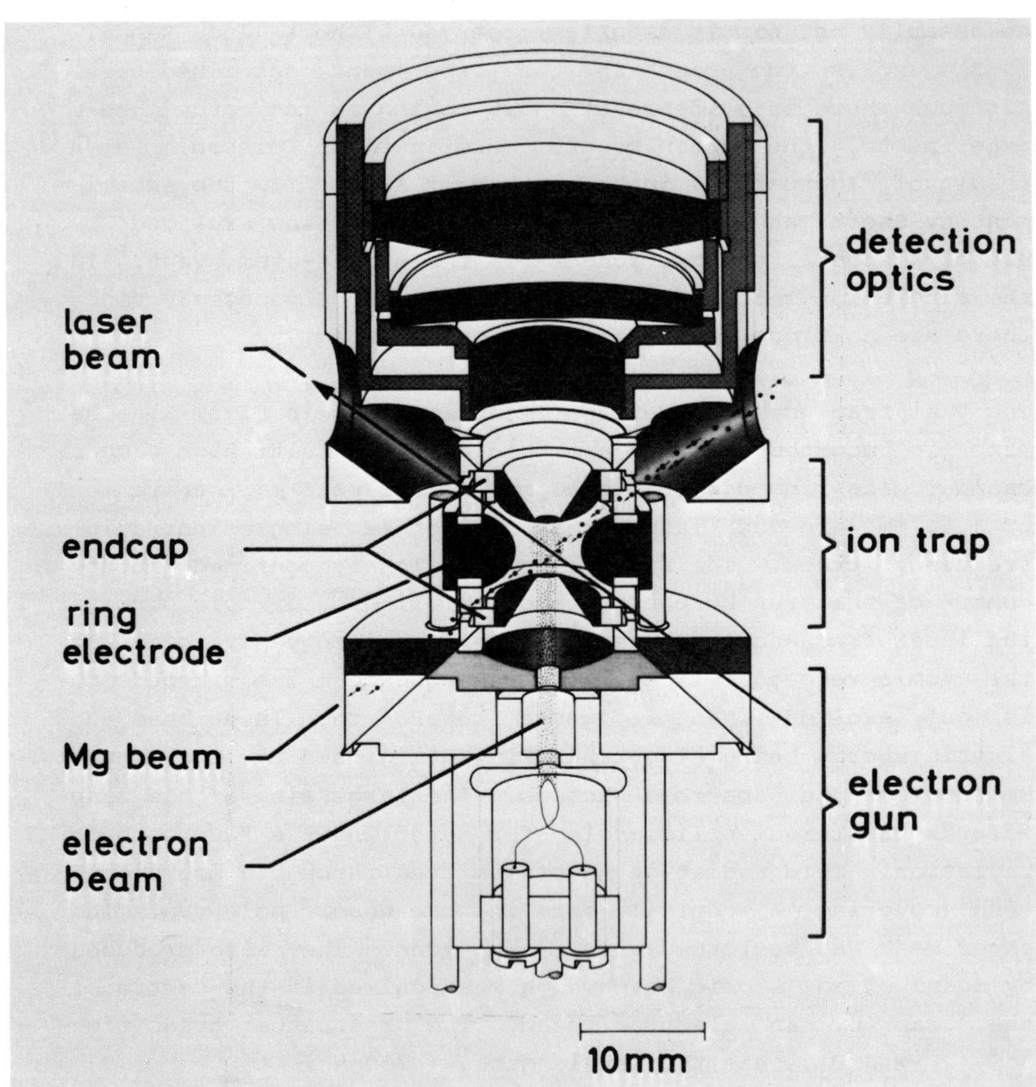

Fig. 3 Scheme of the Paul trap used for the experiments (for details see Diedrich and Walther, 1987)

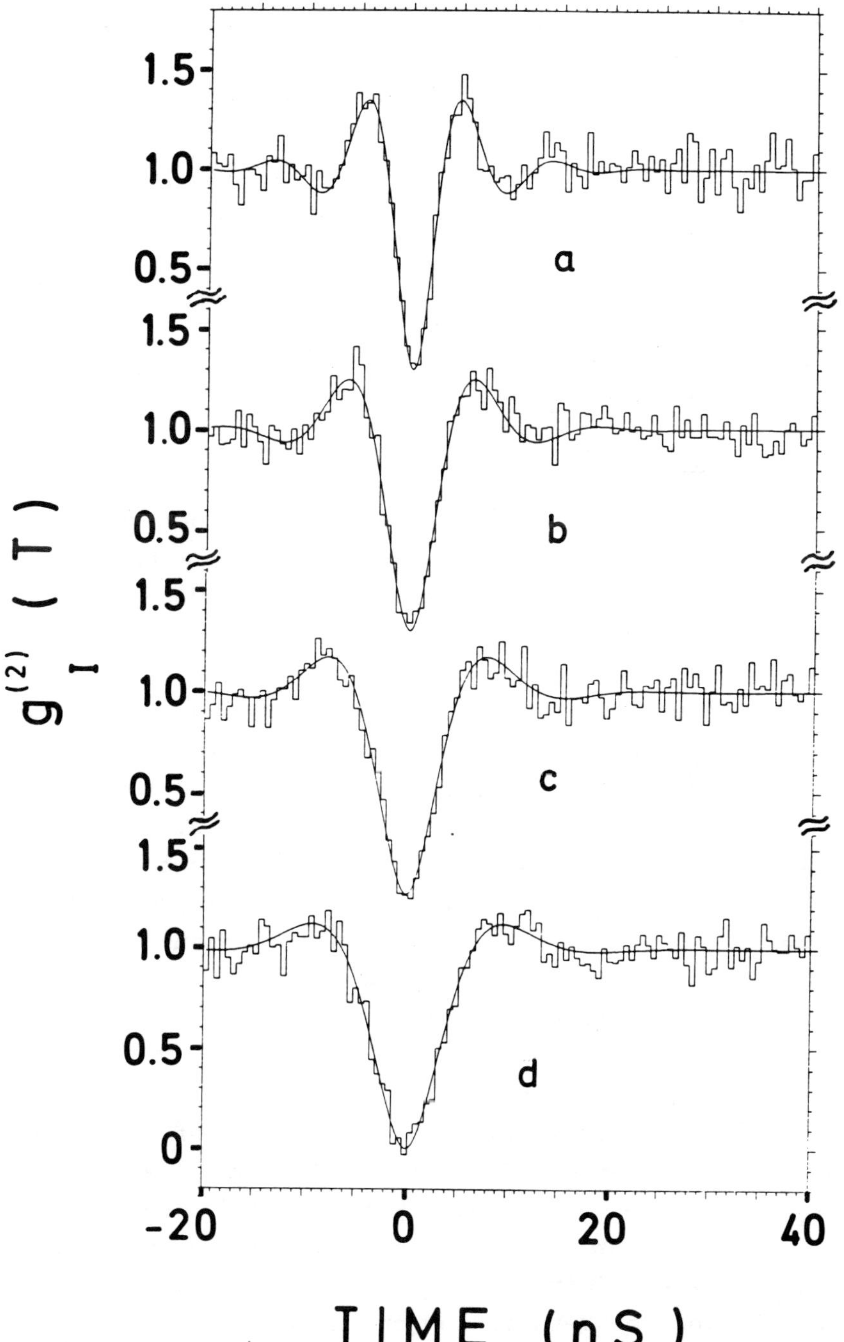

Fig. 4 Results for the intensity correlation. Antibunching for a single ion for different laser intensities, increasing from bottom to top (see Diedrich and Walther, 1987 for details)

be measured for negative τ. The laser intensity decreases from a to d and therefore the average time interval in which a second photon follows a first one increases.

For a single ion the intensity correlation function is given by

$$g^{(2)}(\tau) = 1-e^{-3\gamma\tau/4}\,[\cos\Omega\tau+(3\gamma/4\Omega)\sin\Omega\tau]$$

where $\Omega^2 = \Omega + \Delta^2 - (\gamma/4)^2$, Ω_R being the Rabi frequency at resonance, γ the natural linewidth, and Δ the detuning. In order to check the intensity dependence of Ω^2, a least-squares fit to the data points was calculated. The linear dependence on the intensity could be confirmed. From the signals of Fig. 4 also the lifetime $1/\gamma$ of the excited state could be evaluated. The result agrees quite well with the value known from literature (Diedrich and Walther, 1987).

The fluorescence light of a single stored ion has another interesting property: the fluctuations of the photon number recorded in a small time interval δt is narrower than that expected for a Poissonian distribution, i.e. the variance is smaller than the mean value of the photon number and we again find a sub-Poissonian statistics. The reason is that the single ion can only emit a single photon at a time and fluctuations only occur owing to the finite detection probability (Diedrich and Walther, 1987).

Another interesting phenomenon was recently observed in a Paul trap being connected with laser cooling of the ions. This experiment was performed by Diedrich et al. (1987) with clouds of 2 to about 50 simultaneously stored Mg^+ ions. Two phases of the ions could be clearly distinguished by their excitation spectra. (The excitation of the ions was again performed by coherent tunable UV-radiation on the resonance transition.)

The two phases observed correspond to a cloud-like state, in which the ions move randomly, and to a crystalline state, in which they are fixed in a regular array. The latter state is determined by the pseudopotential of the trap and the Coulomb repulsion of the ions. The crystalline structure corresponds to a Wigner crystal predicted many years ago. Such crystals are formed when laser cooling reduces the kinetic energy of the ions to a value which is much smaller than the potential energy of the ions in the trap.

Measurements with five ions are shown in Fig. 5. The crystalline structure shows a deviation from the pentagon corresponding to the minimum energy configuration for five ions. This is caused by a small contact potential due to a thin Mg coating on the ring electrode of the trap opposite the

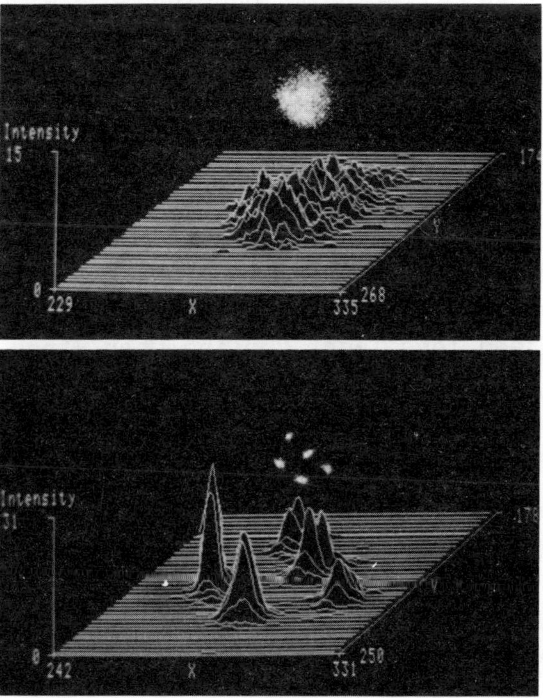

Fig. 5 Observation of five Mg^+ ions in the Paul trap. The structure is observed through a hole in one of the end caps of the trap by using an ultrasensitive camera. The upper figure shows the ions in the cloud phase and the lower one in the crystalline phase. The average distance between the ions is 19μm (Peik, 1987).

Mg oven. The transitions between the crystalline and cloud states were observed by a very sensitive imaging system with single photon detection capability and were also recorded by a video recorder. The transitions between the two phases occur in times as short as 10 ms as determined in separate experiments.

It should be mentioned here that Wigner crystals (as shown for five ions in Fig. 5) are also expected to be seen in ion storage rings with electron cooling now being constructed in many laboratories in the world. In such a storage ring the ions are also transformed from a dilute and weakly coupled Coulomb system into a strongly coupled one with long-range correlations. These beam crystals are very different from ordinary crystals and have lattice spacings similar to that shown in Fig. 5.

The ion crystals in the trap represent a very neat model system. It is of considerable interest to investigate how the conditions depend on the stored ion number: Certain ion configurations can be expected to be more stable than others and will therefore need less laser cooling when they are formed. Furthermore, the vibrational modes of the crystalline structure can be investigated as well as the dynamics of the crystallization and evaporation process.

References

Barton, G. 1987, Proc. Roy. Soc., London A 410 147 and 175

Brune, M., Raimond, J. M., Goy, P., Davidovich, L., Haroche, S., 1987, Phys. Rev. Lett. 59 1899

Carmichael, H. J. and Walls, D.F. 1976, J. Phys. B 9 L43

Cohen-Tannoudji, C. 1977, in Frontiers in Laser-Spectroscopy, ed. by R. Balian, S. Haroche, S. Liberman, North Holland

Cresser, J.D., Häger, J., Leuchs, G., Rateike, M., Walther, H. 1982, in Dissipative Systems in Quantum Optics, ed. by R. Bonifacio, Topics in Current Physics, Vol. 27 21

Springer Verlag
Diedrich, F., Walther, H. 1987a, Phys. Rev. Lett. 58 203
Diedrich, F., Peik, E., Chen, J. M., Quint, W., Walther, H. 1987, Phys. Rev. Lett. 59 2931
Eberly, J. H., Narozhny, N. B., and Sanchez-Mondragon J. J. 1980, Phys. Rev. Lett. 44 1323
Filipowicz, P., Javanainen, J., Meystre, P. 1986, Phys. Rev. A 34 3077
Gallas, J. A. C., Leuchs, G., Walther, H., Figger, H. 1985, in Advances in Atomic and Molecular Physics, Vol 20, eds. D. Bates and B. Bederson, 413, Academic Press 1985
Haroche, S., Raimond, J. M. 1985, in Advances in Atomic and Molecular Physics, Vol 20, eds. D. Bates and B. Bederson, 347, Academic Press
Ihe, W., Anderson, A., Hinds, E. A., Meschede, D., Moi, L., Haroche, S. 1987, Phys. Rev. Lett. 58 666
Jakeman, E., Pike, E. R., Pusey, P. N., and Vaugham, J. M. 1977, J. Phys. A 10 L257
Jaynes, E. T., Cummings, F. W. 1963, Proc. IEEE 51 89
Kimble, H. J., Dagenais, M., and Mandel, L. 1977a, Phys. Rev. Lett. 39 691
Kimble, H. J., Dagenais, M., Mandel, L. 1978b Phys. Rev. A 18 201; Dagenais M., and Mandel, L. 1978, Phys. Rev. A 18 2217
Knight, P. L., and Radmore, P. M. 1982, Phys. Lett. 90A 342
Knight, P. L. 1986, Phys. Scri. T12 51
Krause, J., Scully, M.O., Walther, H. 1987, Phys. Rev. A 36 4547
Lugiato, L. A., Scully, M. O., Walther, H. 1987, Phys. Rev. A 36 740
Meschede, D., Walther, H., Müller, G. 1985, Phys. Rev. Lett. 54 551
Meystre, P. 1987, Opt. Lett. 12 669, and Meystre, P., Wright, E. M., Phys. Rev. A to be published
Narozhny, N. B., Sanchez-Mondragon, J. J., and Eberly, J. H. 1981, Phys. Rev. A 23 236
Peik, E. 1987, Diplomarbeit University of Munich
Rempe, G., Walther, H., Klein, N. 1987, Phys. Rev. Lett. 58 353
Short, R., and Mandel, L. 1983, Phys. Rev. Lett. 51 384; and

in Coherence and Quantum Optics V, edited by Mandel, L. and Wolf, E. 1984, Plenum, New York 671

Yoo, H. I., Sanchez-Mondragon, J. J., and Eberly, J. H. 1981, J. Phys. A 14 1383

Yoo, H. I., and Eberly, J. H. 1985, Phys. Rep. 118 239

PHOTODETECTION AND PHOTOSTATISTICS

M LE BERRE

1. INTRODUCTION

Twenty five years ago, photoelectric detection was simply understood as a conversion of an optical flux into an electrical signal proportional to the beam intensity. This point of view concerns high density optical flux providing c.w. output from the detector.

In the case of an attenuated light beam, the detector output is no more a continuous current ; one can observe successively some bounded electrons making a transition to a continuous state. The emission times are random, even for a constant input intensity. These random times define the photoelectron point process, which is in general the only one available experimentally, the photon point process being inaccessible to direct measurement. One can attempt to derive some information of the field from the measured random properties of the photoelectron process : this is the object of photodetection theory developped first for classical fields by Mandel (1958, 1959, 1963, 1964) and Lamb et Scully (1969). Several quantum treatments have followed (Kelley and Kleiner, 1964 ; Glauber, 1965 ; Meltzer and Mandel, 1964 ; Rousseau, 1975 ; Rousseau, 1977 ; Rocca, 1973 ; Arneodo and Rocca, 1974 ; Kimble and Mandel, 1984). The close correspondence between these two approaches was studied (Sudarshan, 1963 ; Glauber, 1970 ; Klauder, 1966 ; Klauder and Sudarsham, 1968).

More recently a heuristic model of photodetection process, based upon a population monitoring has been proposed which exploits the possible Markov nature of the field evolution (Shepherd, 1981 ; Shepherd, 1984 ; Srinivasan, 1987).

We present first a review of photodetection theory, (classical treatment in section II, quantum theory in section III and population monitoring model in section IV). Then we treat two interesting applications to photostatistics : in section V the photodetection equation is used to classify quantum and classical fields, finally an example of collective antibunching of photons is presented in section VI.

2. PHOTODETECTION CLASSICAL TREATMENT

From a set of photoelectron emission times $\{t_i\}$, one can typically study either the probability distribution of the time intervals between successive events ('intervallometry'), or the probability distribution p(n;T) of obtaining n photocounts in a given interval (t,t+T) ('counting'), or finally the joint probability P_n $\{t_i\}$ of detecting one photoelectron in each subintervals $\{t_i, t_i+\Delta t_i\}$ i= 1...n ('coïncidence' or correlation). The photoelectron process is of course related to the statistical properties of the incident light, as it will be stated.

The first explicit expression for the photocounting distribution p(n;T) was stated by Mandel (1958, 1959, 1963, 1964) who postulate the Poisson compound character of the photoelectron process, since his derivation follows from the two assumptions :

a) The probability of a single photoemission in a short time δt is proportional to the instantaneous classical light intensity I(t), and to δt (there is no accumulation point).

b) Different photoemissions are independent statistical events (the process has no memory, conditionally to a given intensity I(t)).

One can easily prove, as an exercice of probability theory, that the above two hypothesis are necessary and sufficient conditions for a point process to be of Poisson type.

Let us suppose that the intensity I(t) is deterministic, and cut the interval (t,t+T) into N short subintervals of duration δt. The number n of photoelectrons registered between t and t+T can be interpreted as the sum of a large number ($N=T/\delta t$) of independent variables δn_i. Each δn_i equals either to 0 or to 1, registered during the successive subintervals of duration δt

$$n(T) = \sum_{i=1}^{N} \delta n_i \tag{1}$$

The probability that δn_i is equal to unity is $p_i = \eta\, I(t_i)\delta t$, where η is the detector efficiency and $(1-p_i)$ is the probability that δn_i is equal to zero. The characteristic function for n(T) defined as

$$\varphi_n(u) = \langle e^{iu\, n(T)} \rangle \qquad (2)$$

where the brackets means an average over the random values of n(T), can be expressed with Eq.(1)

$$\varphi_n(u) = \lim_{\delta t \to 0} \prod_{i=1}^{N} \left[p_i e^{iu} + (1 - p_i) \right] . \qquad (3)$$

or

$$\log \varphi_n(u) = \lim_{\delta t \to 0} \sum_i \log \{ p_i (e^{iu} - 1) + 1 \}$$

$$= (\eta \int_t^{t+T} I(t')dt')(e^{iu} - 1) \qquad (4)$$

Finally we obtain

$$\varphi_n(u) = \mathrm{Exp}\, \{ \eta\, W\, (e^{iu}) \}\, e^{-\eta W} \qquad (5)$$

where

$$W = \int_t^{t+T} I(t')\, dt' \qquad (6)$$

and by Fourier transform

$$p(n;T) = e^{-\eta W} \frac{(\eta W)^n}{n\,!} \qquad (7)$$

For a random incident light, the distribution in Eq.(7) has to be averaged over the random parameter W, which gives the so-called 'photodetection equation'

$$p(n;T) = \int_0^{\infty} P(W)\, e^{-\eta W} \frac{(\eta W)^n}{n!}\, dW \qquad (8)$$

characteristic of a 'compound Poisson process'.

Averaging over W may change drastically the probability in Eq.(7) which is a sort of bell shape curve. For example in the case of a Gaussian light with coherence time $\tau_c >> T$, $p(W) = 1/W_0 \exp(-W/W_0)$, Eq.(8) leads to the well-known 'Bose Einstein statistics'

$$p(n;T) = (\eta W_0)^n \,/\, (1 + \eta W_0)^{n+1} \qquad (9)$$

which is a monotomic decreasing function (see for example Rousseau, 1969, Fig. 1). Superposition of constant field and Gaussian one was extensively treated (see the bibliography given by Pike, 1970). The first full photocounting distributions reported with modern fast counting experiment were by Johnson et al. (1966).

Let us point out that the photoelectron process has two sources of randomnes :

(a) Even if we consider a constant intensity incident on the detector (coming, for example, from a stabilized monomode laser), the photoelectron process consists in a set of random times $\{ti\}$ randomly distributed without any memory. This very strong property, which will be justified later by quantum theory, defines a pure Poisson process. In that case, (Eq.7), the mean number of photoelectrons is equal to the variance

$$\langle n \rangle = \sigma^2_n = \eta W \qquad (10)$$

(b) If, in addition, the incident intensity fluctuates, as it is the case for usual sources, the average over W, (leading to the compound Poisson process in Eq. (8)), supplies a memory to the process. More precisely the photoelectrons tend to be bunched, as it can be illustrated by the two fold coïncidence probability

$$d^2 / dt_1 dt_2 \, P^{(2)}(t_1 t_2) = \eta^2 < I(t_1) \, I(t_2) > \qquad (11)$$

which has a maximum for $t_1 = t_2$ (Schwartz inequality). This property was first observed in the experiments of Forrester et al. (1955) and of Hanbury-Brown and Twiss (1956, 1957 a-b). This bunching effect is not attributable to the Boson nature of photons, since the laser shows no bunching property whatsoever ; the memory comes from the fluctuations of the incident light.

A consequence of the bunching effect is the enhancement of the fluctuations in a counting experiment. Eq. (10) becomes

$$\sigma^2_n = <n> + \sigma^2_I \, (\eta \, T)^2 \qquad (12)$$

where σ^2_I is the variance of the intensity.

The double random character of the photodetection process prevents a simple correspondence between the statistical properties of the intensity and the statistics of the photocounts. A formal expression for p(W) in terms of p(n;T) and derivatives of Dirac function was proposed (Perina, 1970). However it is more instructive to connect the photoelectron process with the quantum nature of the electromagnetic field. The best way is the use the description of fields in terms of the famous P-representation introduced by Glauber (1963).

Let us consider for simplicity a monomode field (W = I T). The photon distribution or probability of finding N photons in the mode, is the diagonal element of the density matrix in the number representation

$$\rho_{NN} = \frac{1}{N!} \int P(|\alpha|^2) \, |\alpha|^{2N} \, e^{-|\alpha|^2} \, d^2\alpha \qquad (13)$$

Therefore the photocounting distribution p(n;T) derived above for classical fields reproduce the photon number distribution ρ_{NN} *save for the coefficient* $\eta \, T$

Let us precise that the first assumption of Mandel implies that the source, field and detector are supposed to be in a steady state, and consequently the photons absorbed in the detector are continuously replenished by the source.

We must now justify the assumptions leading to Eq.(8), that can only be achieved with a fully quantum mechanical treatment of photodetection.

3. QUANTUM TREATMENTS

The first fully quantum-mechanical treatments of photodetection were given by Kelley and Kleiner (1964) and Glauber (1965). The photodetector is supposed to be constituted by n independent atoms placed at different positions r_1 ... r_n. A shutter in front of all of the atoms is opened during the time interval (t,t+T). Each atom is free to undergo photoabsorption transitions such as the photoelectric effect. Given the initial state of the light field and the atoms, they answer to the question how does the system evolve ? The matrix elements of the operator which induces the transitions are calculated, the probability of detecting one photoabsorption between t, t+T was find after simple sums and averages

$$P^{(1)}(T) = \int_t^{t+T} dt' \int_t^{t+T} dt'' \; S(t''-t') \; G^{(1)}(rt', rt'') \qquad (14)$$

In this expression $G^{(1)}$ is the first order correlation function for the field $G^{(1)}(x_1,x_2) = T_r \{\hat{\rho} \; \hat{E}^-(x_1) \hat{E}^+(x_2)\}$ with $X = (r,t)$, $\hat{\rho}$ is the time independent field density matrix, and S(t) is the response function of the detector, i.e. the Fourier transform of the 'sensitivity' (Glauber, 1965, p. 81). In the limiting case of extremely broadband detectors, the detection process becomes approximately instantaneous in time ; such devices are called ideal photodetectors, $S(t) = \eta \; \delta(t)$, Eq.(15) reduces to

$$P^{(1)}(T) = \eta \int_t^{t+T} G^{(1)}(rt', rt') \, dt' \qquad (15)$$

For ideal photodetectors immersed in the field since time t, the n-fold joint probability density that one count is observed around each of the instant t_i (nothing being specified about the rest of the experiment) is

$$\frac{d^n}{dt_1 \ldots dt_n} P^{(n)}(t_1,\ldots,t_n) = \eta^n \, Tr\,[\,\hat{\rho}\, T_N\, \{\hat{I}(t_1)\ldots\hat{I}(t_n)\}\,] \qquad (16)$$

where $\hat{I} = \int_{detector} \hat{E}^-(r,t)\,\hat{E}^+(r,t)\,d^3r$, $\hat{E}^+$ and $\hat{E}^-$ are respectively the negative and positive frequency parts of the field operator, ρ is the density operator for the field, and T_N indicates that the operators inside the traces { } should be placed in normal order, and in the apex time sequence (E^- are ordered chronologically while E^+ antichronologically).

The probability that n counts are observed in the interval (t;t+T) was derived from Eq.(16) by Kelley and Kleiner (1964)

$$p(n;T) = Tr\,[\hat{\rho}\, T_N\, \{\frac{(\eta\hat{W})^n}{n\,!}\, e^{-\eta\hat{W}}\}\,] \qquad (17)$$

where

$$\hat{W} = \int_t^{t+T} \hat{I}(t')\,dt' \quad . \qquad (17')$$

It leads to

$$\sigma^2{}_n = \langle n\rangle + \eta^2 \langle T_N (\Delta\hat{W})^2\rangle. \qquad (18)$$

This expression in Eqs (17-17'), has obviously a formal analogyy with the classical photodetection equation in Eq.(8). A discussion about its mathematical justification in the frame of quantum point process was made by Srinivas (1977, 1981).

Let us discuss the validity of Eqs.(17-17'). It appears from Eq.(15) that a photoabsorption probability could increase linearly with T instead of being bounded by unity. This unsavoury feature disappears when one departs from first order perturbation theory and takes account of the depletion phenomena due to the photodetector. This treatment was first made by Mollow (1968), then by Scully and Lamb (1969), Rousseau (1977) and Selloni et al. (1977, 1978). The rederived photocount distribution p(n;T) is still given by Eq.(17) but with

$$\hat{W} = \int_t^{t+T} \hat{I}(t') \, e^{-\eta t'} \, dt' \qquad (19)$$

which only agrees with Eq.(17') for $\eta T \ll 1$.

Finally one can conclude that the very intuitive arguments put forwards by Mandel to derive the photodetection equation are nicely justified by the quantum theory.

Apart from the mathematical justification, the quantum theory allows to go further in many directions.

(a) First it obviously describes a lot of fields which has no classical analogue, i.e. no positive P-representation (Glauber, 1963, 1970). The most familiar example of such 'pure quantum' field is the N-photons field, for which any intensity probability distribution can be defined. The counting probability can be derived from Eq. (17), but is more directly obtained from combinatorics (Mollow, 1968 ; Scully and Lamb, 1969 ; Rousseau, 1977).

$$P(n;T) = \binom{N}{n} \xi(T)^n \, (1 - \xi(T))^{N-n} \quad ; \quad \rho_{kk} = \delta_{Nk} \qquad (20)$$

where

$$\xi(T) = 1 - \exp(-\eta T) \qquad (21)$$

is the probability that the photodetector absorbs one photon when immersed in a one-photon field.

More generally, for any monomode field with density matrix elements ρ_{NN}, the photoelectron distribution in Eq.(20) can also be written as

$$p(n;T) = \sum_{N=n} \binom{N}{n} \rho_{NN} \xi(T)^n (1 - \xi(T))^{N-n} \qquad (22)$$

The detector process is therefore shown to progressively attenuate the field towards the vacuum state. For small time sampling, the photon distribution ρ_{NN} is Bernouilli sampled and noticeatly distorted by the detection process. When all the photons are absorbed by the detector ($\eta T >> 1$), the photoelectron distribution is clearly equal to the photon distribution.

Two 'Quantum' effects are now discussed.

1°/ The variance of the photon distribution is $\sigma^2{}_n = < N >$ for a coherent field, $\sigma^2{}_N = < N > + \sigma^2{}_I$ for any random classical field, and may be smaller than $< N >$ for a pure quantum field without any classical analogue. It is the famous 'Sub Poissonian' statistics. Let us define the normalized fluctuation intensity correlation function $(g^2(\tau)-1) = \lambda(\tau)$ or

$$\lambda(\tau) = < T_N (\Delta \hat{I}(t) \Delta \hat{I}(t+\tau) > / < I(t) > < I(t+\tau) > .$$

The variance of the photoelectron can be written as

$$<\Delta n^2> = <n> + \frac{<n>^2}{T} \int_{-T}^{T} (1 - \frac{|\tau|}{T}) \lambda(\tau) d\tau.$$

This expression clearly shows that sub-Poissonian statistics necessarily requires a negative function $\lambda(\tau)$. A trivial example is the monomode N-photon field which has a constant negative $\lambda(\tau)$ and $\sigma^2{}_n = 0$.

The sub-Poissonian character was observed experimentally by Short and Mandel (1963 a-b) in the process of resonance fluorescence emitted from one atom.

2°/ The antibunching effect is another purely quantum effect. It is also due to a non classical behaviour of $\lambda(\tau)$. For classical fields $\lambda(\tau)$ has a bump at $\tau=0$ (Eq. 11), ('bunching effect'). But this property may be inverted, even in the case where λ is positive.

It is the case for photons spontaneously emitted in atomic beam experiment : the 'intensity' correlation of the light is minimum for equal times $t_1 = t_2$ (Kimble, Dagenais and Mandel, 1977 ; Cresser et al., 1982). This photon antibunching effect has a simple quantum explanation : just after the emission of a first photon, the atom is in its ground state and then it cannot immediately emit a second photon. A collective antibunching effect is reported in section VI.

These two quantum effects (sub Poissonian statistics and antibunching) are distinct as it is clearly explained by Short and Mandel (1983-b, p 673).

(b) Moreover one can imagine other sorts of detectors. Mandel (1966), for example, suggested to use the process of stimulated emission as a basis for detection. The effect of shining the light on these spontaneously decaying atoms is that they emit photons rather than absorb them, therefore the operators would occur in anti-normal order in Eq. (17). Different operator orderings are associated with different sorts of experiments. Symmetrical ordering of the operators is also suggested (Glauber, 1970 , p. 72). The problem of ordering field operators and the correspondance between functions of c-numbers and functions of q-numbers has been systematically treated by Agarwal and Wolf (1968 a-b) and Lax (1968).

(c) The above derivations suppose ideal photodetectors with broad band sensitivity function $S(\omega)$, much broader than the spectral width of the incident light. These expressions need to be modified when the optical field has a substantial bandwidth over which the detector response varies (see Eq. 14). Glauber (1965)indicates that the correlation function entering into the expression for $P^{(n)}(t_1...t_n)$ has to be convolved by a certain response function of the detector. This problem was further investigated by Rousseau (1975) and Kimble and Mandel (1984) : a first order perturbation theory shows that the photocounting distribution is still given by Eq.(17), together with

$$\hat{W} = \int\int_t^{t+T} dt' \, dt'' \, \hat{E}^-(t') \, \hat{E}^+(t'') \, S(t''-t') \; . \qquad (23)$$

This result definitely discarts the possibility for the photodetector to count photons for such non resonant interactions between a bound electron and the field (Kimble and Mandel, 1984).

The bunching effect of photoelectrons was investigated for a Gaussian light. While some discrepancy seems to exist between the two papers, it is, in fact, fictitious. The coïncidence probability $P^{(2)}$ for detecting two photoelectrons during the time intervals $(t_1, t_1+\Delta t)$ and $t_2, t_2+\Delta t)$ corresponds to different experiments : Kimble and Mandel treat the case of a detector immersed in the field only during these time intervals ; they found "the regular conclusions consistent with one would expect from Gaussian statistics". Whereas Rousseau considers an experiment where the detector receives the light during a long time interval $(-\infty, +\infty)$, the emitted photoelectrons being registered with the help of some shutter open during small intervals $(t_1, t_1+ \Delta t)$ and $(t_2, t_2 + \Delta T)$; in that case the bunching effect was shown to be much larger. The importance of defining the nature of the measurement performed by the photodetector was recently emphasized.

(d) Srinivas (1977, 1981) has pointed out that most of the so-called quantum approaches to the photocounting problem are not truly quantum mechanical, in the sense that quantities like p(n;T) are, as a rule, calculated on the basis of classical probability theory. A correct quantum treatment must employ the quantum probability theory (Davies, 1976), all the probabilities for a set of events depending on the sequence of experiments. He finally agrees with the well-known 'Glauber-Mandel' formula.

4. POPULATION MONITORING APPROACH

Although the problem of photodetection appears to be satisfactorily solved for free fields interacting only with the detector, Shepherd (1981, 1984) proposed a heuristic model which permits evolution of the detected field during the detection process. This treatment derives from the theoretical population growth model of Shimoda (1957). They have shown that many properties of amplifiers can be deduced from birth, death and immigration model of photons population. The field, placed within a cavity, is assumed to be evolving according to a first-order Markov process. The relevant forward Kolmogorov equation is

$$dP/dt\ (N;t) = \Phi(P(N;t)\ ,\ P(N\pm 1;t)\ ;\ N) \qquad (24)$$

where P(N;t) is the probability that N photons are in the field at time t, and the function Φ specifies the system.

In the detection scheme, the field represented by Eq.(24) is brought into interaction with the detector which then monitors the population. The joint distribution P(n,N;T) that there are N photons in the field at time T and that n photoelectrons have been produced in the interval (O,T) obeys the equation

$$dP/dT\ (n,N;T) = \Phi(P(n,N;T),\ P(n,N\pm 1;T),N) + \eta(N+1)\ P(n-1,N+1;T) - \eta\ N\ P(n,N;T) \qquad (25)$$

which may be transformed for the joint generating function

$$Q(s,z,T) = \ <(1-s)^{N}\ (1-z)^{n}> . \qquad (26)$$

The model of population monitoring for a free field is shown to correspond exactly to the depletion model of Mollow (1968), and others (Scully and Lamb, 1969; Rousseau, 1977; Selloni et al., 1969; Srinivas and Davies, 1981).

In the contrary, it is shown (Shepherd, 1981) that a non-destructive sampling scheme disagrees with classical theory, since it leads to substantially different photoelectron count distribution with a paired ordering of field operators ($(a^{+}a).(a^{+}a)$), or

$$p(n;T) = \ < e^{-\eta T N}\ \frac{(\eta T N)^{n}}{n!} > \qquad (27)$$

noticeably different from Eq.(17) !.

The time evolution of population of photon interacting with the cavity atoms and detector was recently presented in the book of Srinivasan (1987). The markov property for the field implies that the photon population evolves as a branching process. An analysis of the point process of emissions provides the expression for the correlation of photocounts in two intervals and the statistics of dead-time corrected points.

5. INTENSITY MODULATION IN CLASSICAL AND QUANTUM FIELDS

The above description of the evolution in the theory of photodetection is necessary not merely for history but also for a more comprehensive understanding of the phenomena. A first glance to Eqs(17-19) brings immediately the information that intensity and time do not play the same role in a counting experiment, unlike what one would conclude from classical expressions Eqs.(6,8). When increasing the time duration of the experiment, the photoelectron statistics evolves therefore towards the photon statistics

$$\lim_{T\to\infty} P(n;T) = \rho_{nn} \tag{28}$$

However, what happens when the intensity increases ? This problem was studied by Picinbono and Rousseau (1977). Let us present an intuitive outline of their derivation. The intensity of a field is said to be increased or decreased, say 'modulated' by a factor β, (without any consideration for the modulation process), if the new set of intensity momenta $G_\beta{}^{n,n}$ is related to the original one $G^{n,n}$ by the relation

$$G_\beta{}^{n,n}(\{t_i\}) = \beta^n G^{n,n}(\{t_i\}) \tag{29}$$

where

$$G^{n,n}(\{t_i\}) = \mathrm{Tr}\,[\rho\, T_N\, \hat{I}(t_i)....\hat{I}(t_n)\,] \tag{30}$$

For a single mode field with P-representation, the number of photons distribution in the mode (Eq. 13) becomes

$$\rho_{NN}(\beta) = \int_0^\infty p(i)\,.\,e^{-\beta i}\,\frac{(\beta i)^N}{N!}\,di \tag{31}$$

when β increases, the mean intensity of the field and of the number $N(\beta)$ defined by Eq.(31) also increases. But the Poisson function in Eq. (31) becomes relatively norrower and narrower, which allows to show that the scaled variable N_β/β tends to the light intensity

$$P(N_\beta/\beta) \rightarrow P(i) \quad . \tag{32}$$

This relation implies that, among the fields with P-representation, only the classical fields (those with non-negative P function) are 'consistent for modulation', i.e. can be modulated as much as one wants.

The step further was to study fields which have no P-representation. They can be defined via the regularized P-representation studied by Cahill (1969)

$$\rho = \int |\alpha\rangle\langle\alpha| \, P_1 (|\alpha|^2) \, d^2\alpha + \int |-\alpha\rangle\langle\alpha| \, P_2 (|\alpha|^2) \, d^2\alpha \qquad (33)$$

The condition of consistency for modulation is shown to require that $P_2 = 0$, which means that the fields has a P-representation. For example, for the k-photon field ($\rho_{NN} = \delta_{Nk}$), one can construct the modulated field via Eq. 29, it leads to

$$\rho_{NN}(\beta) = \frac{k!}{N!\,(k-N)!} (1-\beta)^{k-N} \beta^N \qquad 0 \leq n \leq k \qquad (34$$

This binomial distribution proves that 'attenuation' of a k-photon field is merely the repeated trial on each photon to be kept with a probability β. However the 'amplification' of a k-photon field, as defined from Eq. 29 with $\beta > 1$, would lead to a negative binomial distribution, and therefore is not possible.

Finally only the quantum fields which have a classical analogue (non-negative P-representation) are shown to be consistent for modulation, while the pure quantum ones cannot be amplified or attenuated without loosing their original statistical properties.

6. ANTIBUNCHING EFFECT IN PHASE-MATCHED RESONANCE FLUORESCENCE

The resonance fluorescence emitted by a single two-level atom was cited in section III-(a) as an example of 'pure quantum' field which has no classical analogue since it exhibits antibunching of photons. This effect was usually left for unrelevant when the source has a large number N of atoms, because the intensity correlation in Eq.(12) was supposed to include N-single-atom antibunching terms and N^2 terms contributing to the usual coïncidences between photons emitted by a chaotic source. However an antibunching of photons emitted by a high atomic density cell was predicted by Le Berre, Ressayre and Tallet (1979). This effect was recently nicely illustrated and observed by Grangier et al. (1986).

Let us briefly explain how phase matching can be achieved to allow some constructive interference from the N atoms.

a) In the work of Le Berre, Ressayre and Tallet, the phase matching arises in the forward fluorescence emitted at the observation point $\vec{r}$ defined by (θ,z), by N atoms supposed to be inside a cylinder of radius a and length l along the z-axis,

$$\hat{E}(r,t) = K \sum_n \hat{D}_n (T_n) / R_n \tag{35}$$

where $R_n = |\vec{r} - \vec{r}_n|$, $T_n = t - R_n/c$, $\vec{r}_n$ is the position of the nth atom and $\hat{D}_n^-$ is the lowering part of the dipole operator for this atom, $K = k^2/4\pi\varepsilon_0$, k is the pump field wave vector.

The fluorescence is isotropic and chaotic, as a result of the central limit theorem, except inside a cone around the Z-axis where there are some constructive interferences : the fluorescence intensity far from the source is the sum of two terms, $I_{inc} + I_{coop}$

$$\hat{I}(r,t) = \frac{K^2}{r^2} \{ \sum_n \hat{D}_n^+ (T_n) \hat{D}_n^- (T_n) + \sum_{n \neq p} \hat{D}_n^+ (T_n) \hat{D}_p^- (T_p) \} \tag{36}$$

The ratio $R(o) = \langle I_{coop} \rangle / \langle I_{inc} \rangle$ is larger than unity inside a cone of half-angle

$$\theta_c \simeq 1.4\ k\ a\ (\bar{N})^{1/3} \tag{37}$$

where $\bar{N}$ is the number of atoms in an absorption length.

The spatial correlation function $G^2(\vec{r},t\ ;\ \vec{r}',t)$ was investigated, it contains

a) N antibunching terms $<\hat{D}^+_p \hat{D}^+_p \hat{D}^-_p \hat{D}^-_p>$.

b) N^2 bunching terms $<\hat{D}^+_n \hat{D}^+_p \hat{D}^-_p \hat{D}^-_n>$, $n \neq p$.

c) The emission of two photons by atom p is also described by antibunching terms, N^2 like $<\hat{D}^+_n \hat{D}^+_n \hat{D}^-_p \hat{D}^-_p>$, $n \neq p$, and N^3 like $<\hat{D}^+_n \hat{D}^+_m \hat{D}^-_p \hat{D}^-_p>$, $n \neq m \neq p$.

d) N^4 terms like $<\hat{D}^+_n \hat{D}^+_m \hat{D}^-_p \hat{D}^-_q>$, $n \neq m \neq p \neq q$.

The first two sets are isotropic, but the second one (b) masks the first one (a) in an usual experiment. The last terms in (c) and (d) are only significant inside the cone of angle θ_c given by Eq.(38). But the main feature of this effect in an usual experimental set up as given in Fig. 1, lies in the fact that the directivity of the cooperative emission (terms in (d)) tends to rule out the antibunching of photons (terms in (c)).

It is shown that the antibunching terms inside the cone θ_c are maximum if the two detectors are located symmetrically with respect to the z-axis, at (θ,z) and $(-\theta,z)$. In that case

$$G^{(2)}(\theta,-\theta) = <I_{inc}>^2 \{1 + 2J_1 (2ka\theta) / 2ka\theta \}^2 + <I_{coop}>^2 \{1 + \eta(\theta) / 2 \ R(\theta) \}^2 - 1/4 <I_{inc}>^2 \quad (38)$$

In Eq.(39) the parameter $\eta(\theta)$ results from the sum of $<\hat{D}^+_n \hat{D}^+_n>$ over the atom located on a section. For example in the case $\Omega_0 T_2 \gg 1$ one obtains

$$\eta \simeq 1 - 2 J_1 (2\theta a\Omega_0/c) \ / \ (2\theta a\Omega_0/c) \quad (39)$$

Ω_0 being the Rabi frequency and T_2 the relaxation time. The antibunching effect therefore vanishes at $\theta = 0$, it is only significant on the domain

$$c \ \Omega_0/a < \theta < \theta_c \quad (40)$$

The normalized intensity correlation function $g^2(\theta,-\theta) = G^{(2)}(\theta,-\theta) / <I(\theta)>^2$ can easily be deduced from Eqs. (39), (40). It may display a spatial antibunching. For example, with the data proposed in the paper of Le Berre et al.(1979), $g^2(\theta,-\theta)$ is given in Fig. 2.

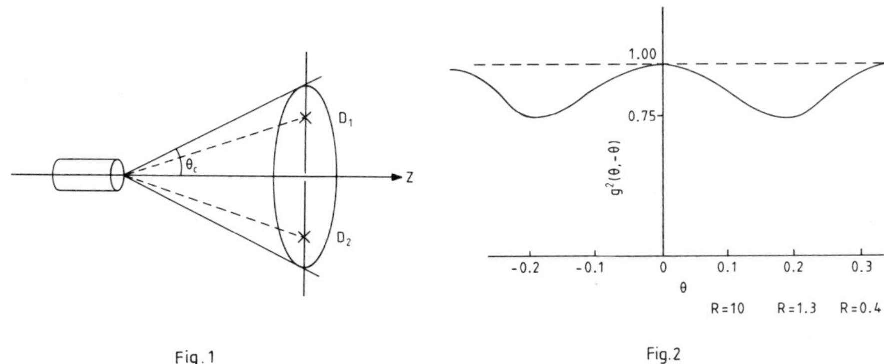

Fig. 1 Fig. 2

In conclusion the antibunching effect can be seen if the following phase matching condition is fulfilled

$$|\vec{k}_p + \vec{k}_p - \vec{k}_1 + \vec{k}_2| \leq 2\,\theta_c\, k \tag{41}$$

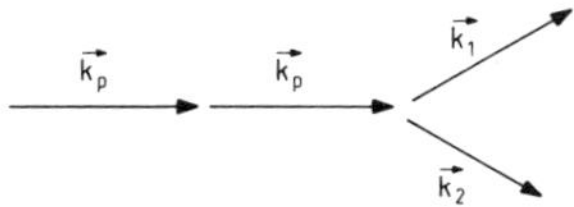

where $\vec{k}_p$ is the laser wave vector, $\vec{k}_1$ and $\vec{k}_2$ the detected field vectors. However the directivity of the cooperative emission cancels the antibunching near $\theta = 0$, it can only be significant in the domain defined by Eq.(40).

(b) In the experiment of Grangier-Roger-Aspect-Heidmann and Reynaud (1986), the competition between the two terms in (c) and (d) is removed. They use the four-wave mixing configuration where the atoms are excited by two counterpropagating pump waves and the fluorescence is observed in two opposite directions (Fig. 3).

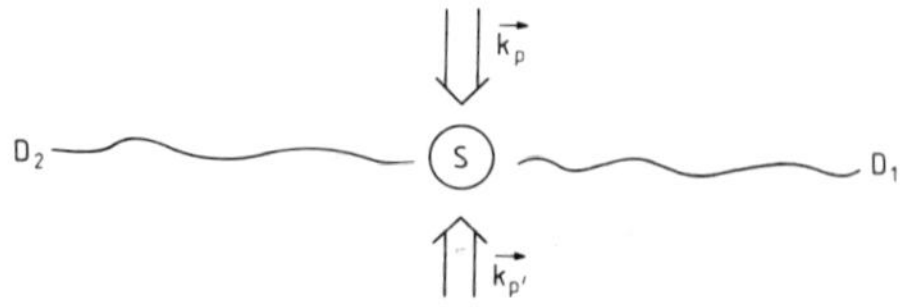

Fig. 3

The antibunching terms therefore scales as N^2, as does the background. The phase matching condition is proved (Heidmann and Reynaud, 1987) to be

$$|\vec{k}_p + \vec{k}_{p'} - \vec{k}_1 + \vec{k}_2| < 1/a \qquad (42)$$

where a is the source radius. The important point in their configuration is that the background intensity is $\langle I_{inc} \rangle$, which gives theoretically zero in the center of the correlation function $g^2{}_{12}(\tau)$. The experimental result reveals a clear antibunching effect, which disappears when the detection misalignment angle is larger than the diffraction angle λ/a which illustrates that phase matching is crucial.

A clear understanding of the correlation properties of terms like $\langle \hat{D}^+{}_n \hat{D}^+{}_n \rangle$ is given in the paper of Heidmann and Reynaud (1987). They propose two approaches, one is an extansion of the fluorescence theory, the second one is based on scattering theory.

REFERENCES

Agarwal, G.S. and Wolf, E. 1968a, Phys. Lett. 26A 485-486 ; 1968b, Phys. Rev. Lett 21 180-183

Arneodo, A. and Rocca, F. 1974, Z. Phys. 269 205-208.

Bedard, G. 1966, Phys. Rev. 151 1038-1039

Bendjaballah, C. 1973, Coherence and Quantum Optics, Proc. Third Conf. on Coherence and Quantum Optics in Rochester, Ed. by L. Mandel and E. Wolf, Plenum Press, New-York, pp 409-414.

Cahill, K.E. 1969, Phys. Rev. 130 1244-1250

Cresser, J.D., Hager, J. Leuchs, G. Rateike, M. and Walther, H. 1982, Dissipative Systems in Quantum Optics, Ed. by R. Bonifacio, Springer-Verlag, New-York.

Davies, E.B. 1976, Quantum Theory of open systems, Academic Press, New York.

Dialetis, D. 1969, J. Phys. A 2 229-235

Forrester, A.T., Gudmundsen, R.A. and Johnson, P.O. 1955, Phys. Rev. 99 1691-1700

Glauber, R.J. 1963a, Phys. Rev. 130 2529-2539 ; 1963b, Phys. Rev. 131 2766-2788 ; 1965, Quantum Optics and Electronics, Ed. by C. deWitt, A. Blandin and C. Cohen-Tannoudji, Gordon and Breach, New-York, pp 63-185.

Glauber, R.J. 1970, Quantum Optics, Proc. Tenth Session of Scottish Universities Summer School in Physics, in 1969, Ed. by S.M. Kay and A. Maitland, Academic Press, London, pp 53-125.

Grangier, P., Roger, G., Aspect, A., Heidmann, A., Reynaud, S. 1986, Phys. Rev. Lett. 57 687-690.

Hanbury-Brown, R. and Twiss, R. . 1956, Nature, London, 177 27 ; 1957a, Proc. Roy. Soc. London A242 300-324 ; 1957b, ibid a243 291-319

Johnson, F.A., MacLean, T.P. and Pike, E.R. 1966, Physics of Quantum Electronics, Ed. Kelley P.L., Lax B. and Tannewald P.E., McGraw-Hill, p. 706.

Kelley, P.C. and Kleiner, W.H. 1964, Phys. Rev. 136, A 316-334.

Klauder, J.R. 1966, Phys. Rev. Lett 16 534-536

Klauder, J.R. and Sudarshan, E.C.G. 1968, Fundamentals of Quantum Optics, Benjamin, New-York.

Kimble, H.J., Dagenais, M. and Mandel L. 1977, Phys. Rev. Lett. 39 691-695

Kimble, H.J. and Mandel, L. 1984, Phys. Rev. A 30, 844-850

Korenman, V. 1967 Phys. Rev. 154, 1233-1240

Lamb, W.E., Scully, Jr. and M.O. 1969, Polarisation Matière et Rayonnement, Presses Universitaires de France, Paris, pp 363

Lax, M. 1968, Phys. Rev. 172, 350-361

Lax, M. and Zwanziger, M. 1973, Phys. Rev. a 7, 750-771

Le Berre, M. Ressayre, E. and Tallet, A. 1979, Phys. Rev. Lett. 43 1314-1317.

Mandel, L. 1958, Proc. Phys. Soc. London 72 1037-1048

Mandel, L. 1959, Proc. Phys. Soc. London 74 233-243

Mandel, L. 1963, Progress in Optics, Ed. by E. Wolf, North-Holland, Amsterdam, vol. II, p. 181.

Mandel, L. Sudarshan, E.C.G. and Wolf, E. 1964, Proc. Phys. Soc. London C 84 435-444.

Mandel, L. 1966, Phys. Rev. 152, 438-451

Meltzer, D. and Mandel, L. 1969, Phys. Rev. 188 198-212

Mollow, B.R. 1968, Phys. Rev. 168 1896-1919

Picinbono, B. and Rousseau, M. Phys. Rev. A 15 1648-1658

Pike, E.R. 1970 Quantum Optics, Proc. Tenth Session of Scottish Universities Summer School in Physics 1969, Ed. S.M. Kay and A. Maitland, Academic Press, London and New York, pp 127-176

Perina,J. 1970, in Quantum Optics, Proc. Tenth Session of Scottish Universities Summer School in Physics, 1969, Ed. S.M. Kay and A. Maitland, Academic Press, London and New York, pp 513-534.

Rocca, F. 1971, Proc. fifth Int. IMEKO Symposium on Photon Detectors, Varna

Rocca, F. 1973, Phys. Rev. D 8 4403-4410
Rousseau, M. 1969a, Journal de Physique 30 675-686
Rousseau, M. 1969b, C.R.Ac. Sc. Paris B 268 1477-1480
Rousseau, M. 1975, J. Phys. A 8 1265-1276
Rousseau, M. 1977, J. Phys. A 10 1043-1047
Scully, M.O. and Lamb, W.E. Jr 1967, Phys. Rev. 159 208-226
Selloni, A. Schwendimann, P. Quattropani, A. and Baltes, H.F. 1977, Optics Comm. 22 131-134; 1978, J. Phys. A 11 1427-1448
Shepherd, T.J. 1981, Optic Acta 28 567-583
Shepherd, T.J. 1984, Optica Acta 31 1399-1407
Short, R. and Mandel, L. 1983a, Phys. Lett. 51 384-386
Short, R. and Mandel, L. 1983b, Proc. fifth Rochester Conference on Coherence and Quantum Optics, Ed. by L. Mandel and E. Wolf, Plenum Press, New York, p. 671-678
Srinivas, M.D. 1977, Jour. Math. Phys. 18 2138-2145
Srinivas, M.D., 1983, Proc. fifth Rochester Conference on Coherence and Quantum Optics, Ed. by L. Mandel and E. Wolf, Plenum Press, New York, p 885-898.
Srinivas, M.D. and Davies, E.B. 1981, Optica Acta 28 981
Srinivasan, S.K. 1987, Point Process Models of Cavity Radiation and Detection, Oxford University Press, New York
Schimoda, K. Takahasi, H. and Townes, C.H. 1957, J. Phys. Soc. Japan 12 686-700
Sudarshan, E.C.G. 1963 Phys. Rev. Lett. 10 277-279

NONCLASSICAL PHOTON INTERFERENCE EFFECTS

C K HONG, Z Y OU AND L MANDEL

1. INTRODUCTION

Since the shortcomings of classical optics in the description of certain fourth-order interference effects were pointed out [Mandel, (1983)], which depend on the joint probability of detecting two photons, there has been interest in the experimental demonstration of these effects. This is in contrast with most second-order interference phenomena, which are well described by classical optics.

Interference occurs in classical optics when two or more wave amplitudes are added with different phases. As a classical particle does not have a phase, only waves can give rise to interference in classical physics. By contrast, interference is a general feature of quantum mechanics, which is not limited to waves, but shows up whenever one can arrive at the outcome of a measurement via several indistinguishable paths, whose probability amplitudes must be added to yield the total probability. Therefore both particle and wave aspects manifest themselves in the quantum mechanical description of interference.

The particle aspect of a photon is apparent in that it cannot be detected at two separate positions at the same time, i.e., the detection of a photon at one point eliminates the probability of detecting the photon at any other point. Therefore Paul (1986) has argued that classical optics, which treats light only in terms of waves, cannot explain some photon interference effects, because a wave can cover an extended region at the same time. Moreover classical optics assumes that a measurement of the light intensity at one point does not affect its value at any other place. Nevertheless, classical optics has successfully described many second-order interference effects.

In this paper we shall discuss two recently reported experiments [Ghosh and Mandel (1987), Hong et al. (1987)] in which the interference between signal and idler photons generated in the process of parametric down-conversion were investigated. Although two-photon interference effects have been treated theoretically before, [for example, Mandel (1983, 1985), Paul (1986)], we will analyze the two experiments in some detail to show how differences between the predictions of classical and quantum theories appear.

2. SPATIAL INTERFERENCE OF TWO PHOTONS

2.1 Background

Until recently, most interference experiments made use of sources, whether correlated or independent, that produce an indefinite number of photons within a coherence time. These cannot exhibit nonclassical interference [Mandel (1983)] with the geometry shown in Fig. 1. Let us consider an experiment in which each of the two light sources A and B emits a single photon. Although there exist some nonclassical interference effects in the resonance fluorescence from single atoms, Ghosh et al. (1986) later showed that the degenerate spontaneous parametric down-conversion process can be used as the source

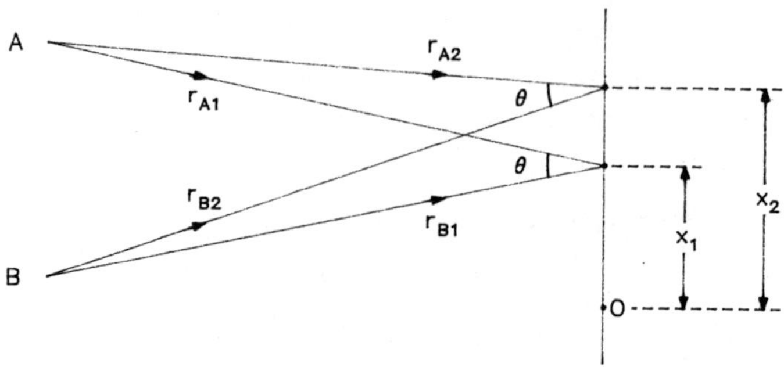

Fig. 1 The geometry of interference experiment with two light sources. [Reproduced from Ghosh, R. and Mandel, L. 1987, Phys. Rev. Lett. 59 1903-1905]

of the two interfering photons. In this process an incoming photon interacting with a nonlinear medium splits into two subharmonic signal and idler photons, whose correlation was experimentally investigated by Burnham and Weinberg (1970), Friberg et al. (1985), and Hong and Mandel (1986).

In the following section the spatial interference between the signal and idler photons of the parametric down-conversion will be treated heuristically, and the predictions of classical and quantum theories will be compared with the experimental results obtained [Ghosh and Mandel (1987)].

2.2 The classical description

In classical optics, the complex analytical signal $V(\underline{r},t)$, which contains only the positive-frequency part of the field, is often used to describe the behavior of the optical field at position $\underline{r}$ and time t. [see Born and Wolf (1980), Mandel and Wolf (1965)]. If we assume that two sources A and B generate fields with the same linear polarization and with the same, well defined single frequency, then the field V_i at position x_i (i = 1 or 2) shown in Fig. 1 can be expressed as

$$V_i = v_A \exp(ikr_{Ai}) + v_B \exp(ikr_{Bi}) , \tag{1}$$

where k is the wave number and v_A and v_B are the complex amplitudes of the fields produced by sources A and B respectively. The time argument has been suppressed, because all the intensities will be evaluated at the same time t. The instantaneous light intensity I_i defined by $I_i = V_i^* V_i$ is easily calculated from Eq. (1), and is given by

$$I_i = v_A^* v_A + v_B^* v_B + v_A v_B^* \exp[ik(r_{Ai}\ r_{Bi})] + c.c. , \tag{2}$$

in which the last two terms exhibit second-order interference. If we assume that the phases of the two sources are random and unrelated, then after we average over the ensemble of all realizations the interference terms disappear, leaving only the constant terms,

$$\langle I_i \rangle = \langle I_A \rangle + \langle I_B \rangle \tag{3}$$

where $\langle I_A \rangle = \langle v_A^* v_A \rangle$, $\langle I_B \rangle = \langle v_B^* v_B \rangle$. Nevertheless interference effects appear in the correlation between the light intensities at two neighboring points.

If we calculate the average of the intensity product I_1I_2 under the same assumption, we obtain

$$\langle I_1I_2\rangle = \langle I_A^2\rangle + \langle I_B^2\rangle + 2\langle I_AI_B\rangle + 2\langle I_AI_B\rangle\cos[2\pi(x_1-x_2)/L]. \quad (4)$$

$L \simeq \lambda/\theta$ is the spacing of the classical interference fringes corresponding to the geometry of Fig. 1, i.e., two waves of wavelength λ coming together at a small angle θ. We note that interference effects show up as a periodic variation of $\langle I_1I_2\rangle$ with $|x_1-x_2|$. The ratio η of the magnitude of the interference term to the constant term, which might be called "visibility" because it is also the relative depth of the modulation, is always less than or equal to 50 %,

$$\eta = \frac{2\langle I_AI_B\rangle}{\langle I_A^2\rangle + \langle I_B^2\rangle + 2\langle I_AI_B\rangle} \leqq \frac{1}{2} \quad (5)$$

As a result, the intensity correlation $\langle I_AI_B\rangle$ does not vanish for any value of $|x_1-x_2|$.

The fact that η cannot be unity in classical wave optics is a consequency of the fact that $\langle I_A^2\rangle$ and $\langle I_B^2\rangle$ can be zero only for fields that vanish identically. This is so because waves can be everywhere at the same time, and the field at one point in general is not affected by a measurement at other point. This is not true in the quantum mechanical description of the interference effect.

2.3 The quantum mechanical description

As practically all detectors of light rely on the photoelectric effect, in which a free electron is produced at the expense of a photon, the photon annihilation operator plays a key role in describing the measurement of light in the quantum theory. If we consider only a single-frequency contribution of each source as in the classical description, then we can express the positive-frequency part of the field, $\hat{E}^{(+)}(x_i)$ at the point x_i in Fig. 1 in the form (Hilbert-space operators are distinguished by a caret)

$$\hat{E}^{(+)}(x_i) = \hat{a}_A\exp(ikr_{Ai}) + \hat{a}_B\exp(ikr_{Bi}), \quad (6)$$

where $\hat{a}_A$ and $\hat{a}_B$ are photon annihilation operators acting on the fields generated by sources A and B respectively. It then follows that the probability of detecting a photon at x_i within some narrow range δx_i is given by [Glauber (1963a,b)]

$$P_1(x_i)\delta x_i = K_i \delta x_i \langle \hat{E}^{(-)}(x_i)\hat{E}^{(+)}(x_i)\rangle , \tag{7}$$

where K_i is a factor characteristic of the detector positioned at x_i, and $\langle\ \rangle$ represents the expectation value of the operator for the given state. If we suppose that A and B correspond to the signal and idler modes of the degenerate parametric down-conversion process, then the appropriate state is the two-photon Fock state $|1_A, 1_B\rangle$, because signal and idler photons are always produced together. For this state and the field operators given by Eq. (6), the expectation in Eq. (7) is easily evaluated and the probability reduces to

$$P_1(x_i)\delta x_i = 2K_i \delta x_i , \tag{8}$$

so that there is no interference term, just as in the classical calculation of the intensity.

However fourth-order interference effects show up in the joint probability $P_{12}(x_1, x_2)\delta x_1 \delta x_2$ of detecting photons at x_1 and x_2 within δx_1 and δx_2, in which case [Glauber (1963a,b)]

$$\begin{aligned} P_{12}(x_1,x_2)\delta x_1 \delta x_2 &= K_1 K_2 \delta x_1 \delta x_2 \langle \hat{E}^{(-)}(x_1)\hat{E}^{(-)}(x_2)\hat{E}^{(+)}(x_2) \\ &\quad \times \hat{E}^{(+)}(x_1)\rangle \\ &= K_1 K_2 \delta x_1 \delta x_2 \Big(\langle \hat{a}_A^{\dagger 2}\hat{a}_A^2\rangle + \langle \hat{a}_B^{\dagger 2}\hat{a}_B^2\rangle \\ &\quad + 2\langle \hat{a}_A^{\dagger}\hat{a}_B^{\dagger}\hat{a}_B\hat{a}_A\rangle\{1 + \cos[2\pi(x_1 - x_2)/L]\}\Big) . \end{aligned} \tag{9}$$

For the two-photon state $|1_A, 1_B\rangle$ the first two terms on the right vanish and we obtain

$$P_{12}(x_1,x_2)\delta x_1 \delta x_2 = 2K_1 K_2 \delta x_1 \delta x_2 \{1 + \cos[2\pi(x_1 - x_2)/L]\}. \tag{10}$$

Interference shows up again as a cosine modulation, as in the classical description, but in this case the visibility η is 100 %. This implies that the detection of a photon at one position can completely rule out the possibility of detecting another photon at certain positions.

It should be noticed that the 100 % visibility is due to the particle nature of a photon, which cannot be detected at two separate positions at the same time. This particle nature appears in quantum mechanics formally in the change of the state of the field from a single-photon Fock state to the vacuum state under the action of an annihilation operator representing the detection of a photon.

In practice, photons are not detected at a point but over extended regions Δx's centered at x_1 and x_2, so that we ought to integrate the foregoing expression over Δx. From Eq. (10), the measured joint detection probability, $P_{12}(x_1,x_2)$, is given by

$$\begin{aligned} P_{12}(x_1,x_2) &= \int_{x_1-\Delta x/2}^{x_1+\Delta x/2}\int_{x_2-\Delta x/2}^{x_2+\Delta x/2} \mathcal{P}_{12}(x_1,x_2)\,dx_1 dx_2 \\ &= 2K_1K_2(\Delta x)^2\{1 + \left(\frac{\sin\pi\Delta x/L}{\pi\Delta x/L}\right)^2 \cos[2\pi(x_1-x_2)/L]\}. \end{aligned} \tag{11}$$

Due to the finite detector width Δx, the visibility η has been reduced by the factor $[\sin(\pi\Delta x/L)/(\pi\Delta x/L)]^2$, and the same factor is applicable to the classical calculation too. In practice this factor was 0.55 in the experiment described below.

2.4 The experiment

The outline of the interference experiment [Ghosh and Mandel

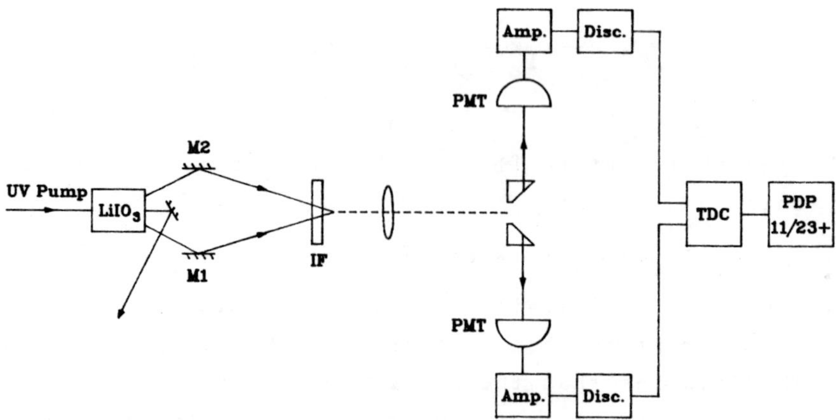

Fig. 2 Outline of the two-photon interference experiment. [reproduced from Ghosh, R. and Mandel, L. 1987, Phys. Rev. Lett. 59 1903-1905]

(1987)] is shown in Fig. 2. Light from an argon-ion laser oscillating or the 351.1 nm UV line falls on a 1.5 cm long $LiIO_3$ crystal, where some UV photons split into half-frequency signal and idler photons, which emerge at angles of ±3.3° to the UV beam. A beam stop deflects the latter, while two mirrors M1 and M2 cause the signal and idler photons to come together at an angle $\theta \simeq 2°$ in a plane 1.1 m from the crystal, after passing through an interference filter with 3×10^{13} rad/s bandwidth. The interference pattern formed in this plane is magnified and re-imaged by a lens so as to make the fringe spacing $L \simeq 0.34$ mm. Two glass plates, each of thickness $\Delta x \simeq 0.14$ mm and mounted on a translator, collect the incoming photons at x_1 and x_2 edge on, and direct them to two photomultiplier tubes, whose pulses, after amplification and shaping, are fed to the start and stop inputs of a time-to-digital converter. Pulse pairs arriving within a 5 ns interval are treated as coincidence counts, due either to the 'simultaneous' signal and idler photons or to the accidental overlap of uncorrelated pulses. When the latter are subtracted out, the rate of coincidence counting provides a

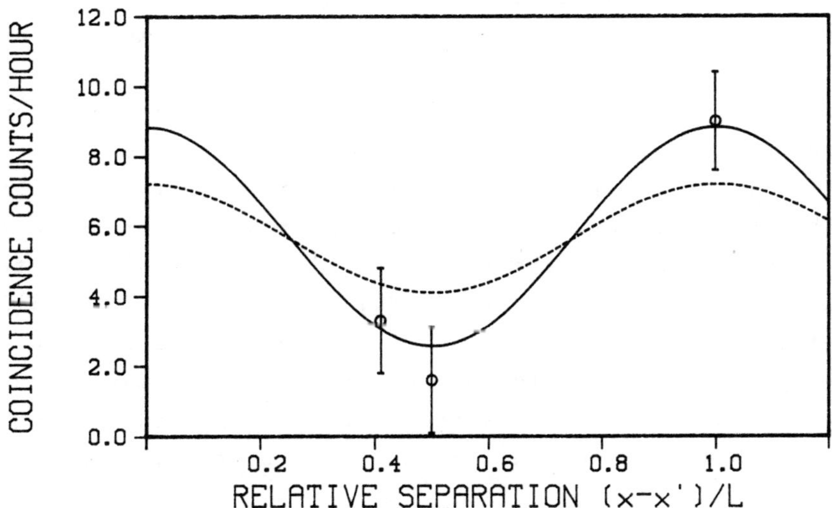

Fig. 3 Experimental results superimposed on the predictions of quantum theory given by Eq. (11) (solid curve), and of the classical theory with maximum modulation (dashed curve). [reproduced from Ghosh, R. and Mandel, L. 1987, Phys. Rev. Lett. 59 1903-1905]

measure of the joint probability $P_{12}(x_1,x_2)$, apart from a scale factor.

Fig. 3 shows the results obtained for three different values of $|x_1-x_2|$. The solid curve in Fig. 3 is a plot of $P_{12}(x_1,x_2)$ given by Eq. (11) with $\Delta x \simeq 0.14$ mm and $L \simeq 0.34$ mm, with the scale factor K_1K_2 adjusted arbitrarily for best agreement with the data. The dashed curve in Fig. 3 is the corresponding prediction by the classical wave theory with maximum visibility of 0.275 for the same Δx and L. The values of χ^2 corresponding to the agreement between the measured values and the two theoretical predictions are found to be $\chi^2_{class} \simeq 4.9$ and $\chi^2_{qu} \simeq 0.44$. The probabilities of obtaining values of χ^2 of this size are $P(\chi^2 \geqq 4.9) \simeq 0.18$ and $P(\chi^2 \geqq 0.44) \simeq 0.92$, respectively. The experimental results therefore clearly favor the prediction of the quantum theory.

3. MEASUREMENT OF TIME CORRELATION BETWEEN TWO PHOTONS

3.1 Background

The time correlation between signal and idler photons produced in the parametric down-conversion process was first measured by Burnham and Weinberg (1970) with a resolving time of about 10 ns. The measurement were repeated [Friberg et al. (1985)] with improved resolving time of order 100 ps. Both experiments depended on measurement of the time intervals between the photoelectric pulses resulting from detection of the two photons, and therefore provided only an upper bound for the correlation time, limited by the resolving time of the photo-detectors. Later Abram and coworkers (1986) measured the correlation time between signal and idler beams by use of intense light pulses, by making use of the process of second-harmonic generation. What they measured was therefore not so much the correlation time between two photons but two intense pulses.

In the next section, we will discuss the experiment carried out by Hong et al. (1987), which not only yielded a subpicosecond correlation time, but also exhibited nonclassical interference effects. The experiment depended on measuring the joint probability of detecting signal and idler photons produced in

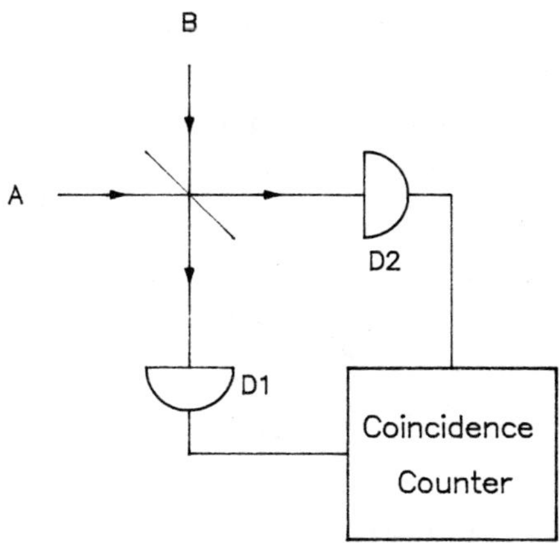

Fig. 4 The schematic of the time correlation experiment.

parametric down-conversion after the down-converted light fields are mixed by a beam splitter, as shown in Fig. 4.

3.2 The classical description

Let us suppose that each source in Fig. 4 emits a very short pulse of identical shape and duration at the same time, and that $v_A g(t)$ and $v_B g(t)$ are the fields from sources A and B. v_A and v_B are constant complex amplitudes, and $g(t)$ is the real pulse envelope corresponding to a bandwidth $\Delta\omega$. We shall assume that $g(t)$ is normalized so that $\int g^2(t)dt = 1$. The phases of the two fields are again assumed to be random and unrelated. Then the combined field V_i falling on the detector i at time t can be expressed as

$$\begin{aligned} V_1(t) &= \sqrt{T}v_A g(t) + i\sqrt{R}v_B g(t+\tau), \\ V_2(t) &= i\sqrt{R}v_A g(t) + \sqrt{T}v_B g(t+\tau), \end{aligned} \tag{12}$$

where R and T are the reflectivity and transmissivity of the beam splitter, with $R + T = 1$, and τ is the difference in the arrival times of the two pulses at the beam splitter. The position argument has been ignored, because the phase difference across the crosssection of the mixed beam is constant.

In Sec. 2 we ignored any consideration of the pulse widths and resolving times of the detectors, because they were irrelevant to the description of the interference effects discussed. But, in practice, what is measured by a detector is not the instantaneous intensity of the field, but its integral over the resolving time Δt, which is assumed to be much greater than the width $\delta t \simeq 1/\Delta\omega$ of $g(t)$. Then the integrated intensities measured by detectors 1 and 2 and their cross-correlation $C(\tau)$ can be calculated as in Sec. 2. We obtain from Eq. (12)

$$\begin{aligned} \int \langle I_1(t)\rangle dt &= T\langle I_A\rangle + R\langle I_B\rangle \\ \int \langle I_2(t)\rangle dt &= R\langle I_A\rangle + T\langle I_B\rangle \end{aligned} \qquad (13)$$

and

$$\begin{aligned} C(\tau) &= \iint \langle I_1(t) I_2(t')\rangle dt dt' \\ &= TR\langle I_A^2\rangle + TR\langle I_B^2\rangle + T^2\langle I_A I_B\rangle + R^2\langle I_A I_B\rangle \\ &\quad - 2TRh(\tau)\langle I_A I_B\rangle, \end{aligned} \qquad (14)$$

where $h(\tau) = [\int g(t)g(t+\tau)dt]^2 \leqq 1$.

It is clear from Eqs. (13) that no interference effects show up in the average integrated intensity. Although Eq. (14) exhibits no periodic variation with τ, the last term, which depends on τ, is an interference term.

If we define visibility η in the usual way as the ratio of the interference term to the constant term, then it follows from Eq. (14) that η is always less than or equal to 50 %, for

$$\eta = \frac{2TRh(\tau)\langle I_A I_B\rangle}{TR\langle I_A^2\rangle + TR\langle I_B^2\rangle + (T^2+R^2)\langle I_A I_B\rangle} \leqq \frac{h(\tau)}{2} \leqq \frac{1}{2}. \qquad (15)$$

The physical reason behind this is the same as discussed in the previous example in Sec. 2.

3.3 The quantum mechanical description

Although we earlier treated the signal and idler modes of the parametric down-conversion process as having well-defined frequencies, the photons actually have a wide bandwidth $\Delta\omega$, and only the sum of the two frequencies is well defined. The two-photon state produced in the down-conversion process can be represented by the linear superposition

$$|\Psi\rangle = \int d\omega\, \phi(\omega_0/2+\omega, \omega_0/2-\omega)\, |1_{\omega_0/2+\omega}, 1_{\omega_0/2-\omega}\rangle, \qquad (16)$$

where ω_0 is the frequency of the pump beam and $\phi(\omega_0/2+\omega,\omega_0/2-\omega)$ is some symmetric weight function which is peaked at $\omega = 0$.

The fields $\hat{E}_1^{(+)}(t)$ and $\hat{E}_2^{(+)}(t)$ past the beam splitter falling on detectors 1 and 2, are related to the signal and idler fields by the following expressions,

$$\begin{aligned}\hat{E}_1^{(+)}(t) &= \sqrt{T}\hat{E}_A^{(+)}(t) + i\sqrt{R}\hat{E}_B^{(+)}(t+\tau)\\ \hat{E}_2^{(+)}(t) &= i\sqrt{R}\hat{E}_A^{(+)}(t) + \sqrt{T}\hat{E}_B^{(+)}(t+\tau)\end{aligned} \tag{17}$$

as in the classical description.

The joint probability of detecting photons by both detectors within the resolving time t, which is much longer than any correlation time of the light, is given by

$$P_{12}(\tau) = K\int\int\langle\hat{E}_1^{(-)}(t)\hat{E}_2^{(-)}(t')\hat{E}_2^{(+)}(t')\hat{E}_1^{(+)}(t)\rangle dtdt' \tag{18}$$

and can be readily calculated from Eqs. (16) and (17). We find [Hong et al. (1987)]

$$P_{12}(\tau) = C[T^2 + R^2 - 2TRh(\tau)] \tag{19}$$

where C is another constant,

$$h(\tau) = \int g(t)g(t+\tau)dt/\int g^2(t)dt, \tag{20}$$

and

$$g(t) = \int\phi(\omega_0/2+\omega,\omega_0/2-\omega)\exp(-i\omega t)d\omega/\int\phi(\omega_0/2+\omega,\omega_0/2-\omega)d\omega. \tag{21}$$

Eq. (19) shows the effects of interference on the variation of $P_{12}(\tau)$ with the time delay τ, as in the classical description (cf. Eq. (14)). However the absence of constant terms proportional to TR in the quantum mechanical result makes it possible for the visibility η to be as large as 100 % for $T = R = 1/2$. Thus from Eq. (19)

$$\eta = \frac{2TRh(\tau)}{T^2 + R^2} \leq h(\tau) \leq 1. \tag{22}$$

The physical reason for the absence of constant terms proportional to TR is the particle nature of the photon, as discussed previously.

The variation of the interference effect with the time delay τ makes it possible to measure $h(\tau)$ and therefore to determine the spread function $g(t)$ of the photons. In the special case, when $\phi(\omega_0/2+\omega,\omega_0/2-\omega)$ is Gaussian in ω with bandwidth $\Delta\omega$, g(t) also has the Gaussian form

$$g(t) = \exp[-(t\Delta\omega)^2/2] , \tag{23}$$

and the expected number of coincidence counts N_c within some measurement time is given by

$$N_c = C(T^2+R^2)\{1 - \frac{2TR}{T^2+R^2}\exp[-(\tau\Delta\omega)^2]. \tag{24}$$

3.4 The experiment

Fig. 5 shows an outline of the set-up used by Hong et al. (1987), which is basically similar to the one used by Ghosh and Mandel (1987). The main differences are the use of a KDP crystal

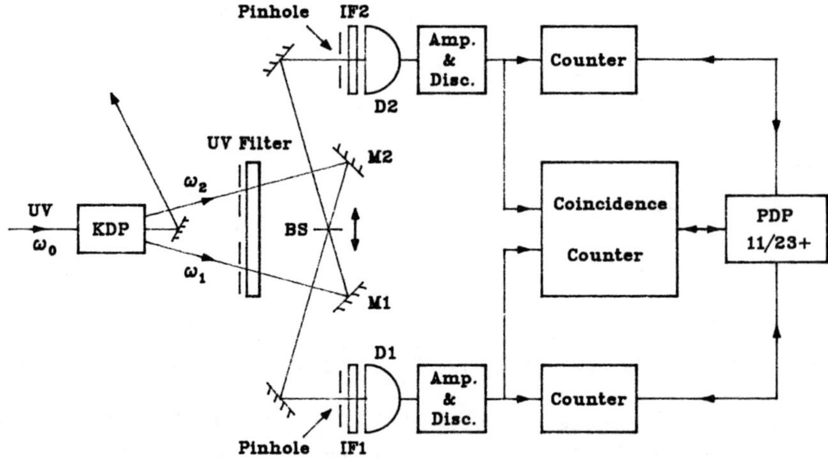

Fig. 5 Outline of the time correlation experiment set-up. [Reproduced from Hong, C. K., Ou, Z. Y., and Mandel, L. 1987, Phys. Rev. Lett. 59 2044-2046]

in place of $LiIO_3$ and the addition of a beam splitter BS to combine the signal and idler beams. The angles of mirrors M1 and M2 are adjusted so that the two beams overlap as well as possible in crosssection, with the same propagation direction, after passing through BS. In order to introduce a time delay between signal and idler fields the beam splitter BS is translated perpendicularly to its face. Pulses received from the two sets of photomultiplier-amplifier-discriminator combinations within a 7.5 ns interval are treated as coincident, and after subtraction of accidentals, yield a measure of the probability $P_{12}(\tau)$.

Fig. 6 is the plot of the coincidence counting rates for several displacements of the beam splitter BS shown in Fig. 5. The fact that the coincidence rate at the center of the dip in Fig. 6 is only a few percent of its value in the wings, whereas it cannot be less than 50 % according to classical theory, demonstrates the nonclassical nature of the interference effect.

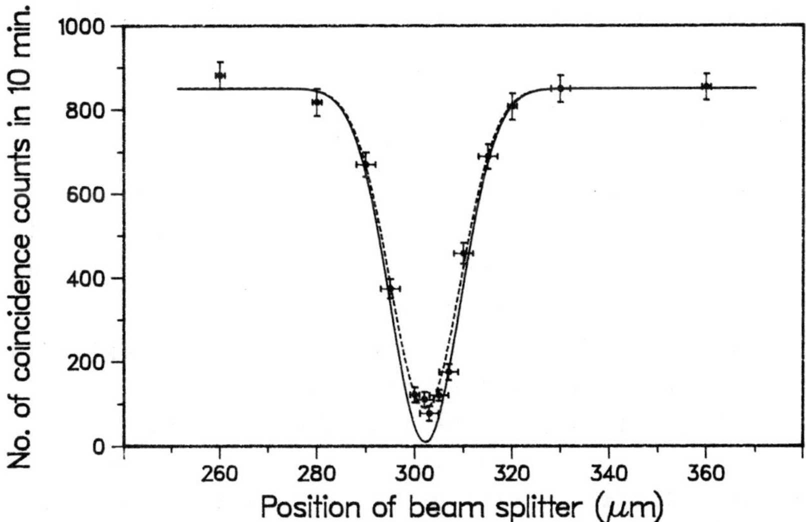

Fig. 6 The measured number of coincidences as a function of beam splitter displacement $c\tau/2$, superimposed on the theoretical curve derived from Eq. (24) with $R/T=0.95$, $\Delta\omega=3\times10^{13}$ rad/s. For the dashed curve the factor $2TR/(T^2+R^2)$ in Eq. (24) was multiplied by 0.9. The vertical error bars correspond to one standard deviation, whereas horizontal error bars are based on estimates of the measurement accuracy. [Reproduced from Hong, C. K., Ou, Z. Y., and Mandel, L. 1987, Phys. Rev. Lett. 59 2044-2046]

The correlation time between the two photons, which is of the same order as the length of the photon wave packet, is found to be 100 fs, which is obtained by doubling the width of the dip at half height. The doubling is necessary because the mirror image travels twice as far as the beam splitter when the latter is displaced. This time is about what would be expected from the passband of the interference filter, which limits the bandwidths of the detected signal and idler photons to 3×10^{13} rad/s.

The solid curve in Fig. 6 is based on the quantum mechanical prediction given by Eq. (24) with the measured values of $R/T \simeq 0.95$ and $\Delta\omega \simeq 3\times10^{13}$ rad/s $\simeq 5\times10^{12}$ Hz, and shows that the coincidence count rate should fall close to zero at the center of the dip. The failure of the experimental points to fall to zero is probably due to a slight lack of overlap of signal and idler beams, causing less than perfect destructive interference. The dashed curve in Fig. 6, obtained by multiplying the factor $2TR/(T^2+R^2)$ in Eq. (24) by 0.9, to allow for the less than perfect crosssectional overlap, matches the experimental points quite well.

4. CONCLUSION

We have therefore demonstrated two examples of a nonclassical effect. In both cases the observed visibility violates classical probability. In both experiments a certain joint detection probability becomes zero or small, because of the destructive interference of two-photon detection probability amplitudes, which are quantities without counterpart in classical optics. Finally, we have shown that the interference effect can be made the basis of a measurement of the length of a photon wave packet with femtosecond time resolution. Even that is by no means the limit. By mounting the beam splitter on a piezoelectric transducer in the second experiment, one should be able to achieve reproducibility and resolution well below a femtosecond.

REFERENCES

Abram, I., Raj, R. K., Ouder, J. L., and Dolique, G. 1986, Phys. Rev. Lett. 57 2516-2519

Born, M. and Wolf, E. 1980, Principles of Optics, 6th ed., Pergamon Press, Oxford, pp 494-497

Burnham, D. C. and Weinberg, D. L. 1970, Phys. Rev. Lett. 25 84-87

Friberg, S., Hong, C. K., and Mandel, L. 1985, Phys. Rev. Lett. 54 2011-2013

Ghosh, R. and Mandel, L. 1987, Phys. Rev. Lett. 59 1903-1905

Ghosh, R., Hong, C.K., Ou, Z. Y., and Mandel, L. 1986, Phys. Rev. A 34 3962-3968

Glauber, R. J. 1963a, Phys. Rev. 130 2529-2539

Glauber, R. J. 1963b, Phys. Rev. 131 2766-2788

Hong, C. K. and Mandel, L. 1986, Phys. Rev. Lett. 56 58-60
Hong, C. K., Ou, Z. Y., and Mandel, L. 1987, Phys. Rev. Lett. 59 2044-2046
Mandel, L. 1983, Phys. Rev. A 28 929-943
Mandel, L. 1985, Open Questions in Quantum Physics, Eds. Tarozzi, G. and van der Merwe, A., D. Reidel Publishing Co., Dordrecht, pp 333-343
Mandel, L. and Wolf, E. 1965, Rev. Mod. Phys. 37 231-287
Paul, H. 1986, Rev. Mod. Phys. 58 209-231

ACKNOWLEDGMENT

This work was supported by the National Science Foundation and by the Office of Naval Research. One of us (C.K.H.) was the recipient of a Newport Research Award.

PHOTONS AND APPROXIMATE LOCALISABILITY

E R PIKE AND S SARKAR

1. INTRODUCTION

It is intuitive to consider particles as localised in space and so, to a good approximation, to have an absolute "position" which does not have to be calculated in some mean sense. (This approximation improves if the particle becomes more and more like a point.) In quantum mechanics conventionally the notion of position is replaced by that of a position operator. Somewhat less conventionally (Mackey, 1978) it is possible to state the same thing by introducing projection operators P associated with each and every elementary (open) cube in three dimensional space. Acting on a wavefunction of the particle such an operator projects out that part of the wavefunction with support in its particular cube. The square of the modulus of this projected wavefunction gives the probability of finding the particle in that cube. The projection is such that this probability does not change as the co-ordinate system of the cube is rotated or translated. For the photon no such squared modulus of a projection with the interpretation of a probability exists (Newton, and Wigner, 1949; Wightman, 1962; Jauch, and Piron, 1967; Pike, and Sarkar, 1986). As is shown in these references, this is the case theoretically because it is a spin one particle with a (single) constraint on the spin degrees of freedom. However, experimentalists are detecting many photons every day in laboratories throughout the world and one thinks of these photons in some sense as point-like. Photons are absorbed and in some sense at fixed instants cause electrons in the detector to jump from a ground to an excited state.

Glauber (1963) produced a theory of photodetection in the so-called "rotating wave approximation" which is usually an acceptable simplification for optical photons. In considering such delicate issues as the localisability of photons it may be wise to proceed initially without making such an assumption. Hence we will calculate the photodetection probability without making this approximation before discussing localisation. In Glauber's model of a one-atom photon detector, the initial state (at t=0) of the atom and field is denoted by $|gi\rangle$, where g is the ground state of the atom and i is some arbitrary initial field configuration. The field is quantised in the Coulomb gauge. If at time t the joint state of the atom field system is $|af\rangle$, where a is the state of the atom and f that of the field, then the probability for this to have happened is, in the dipole approximation, given by

$$\text{Probability of transition} = \left| \frac{e}{i\hbar} \int_0^t dt' \langle af| \sum_j \underline{r}_j(t') \cdot \underline{E}(\underline{x},t') |gi\rangle \right|^2 \tag{1}$$

Here $\underline{r}_j$ is the position operator of the jth electron with respect to the nucleus (at position $\underline{x}$). $\underline{E}$ is the operator for the electromagnetic field at $\underline{x}$. We will write

$$\langle a| \sum_j \underline{r}_j(t') |g\rangle = \underline{D}_{ag}\, e^{i\omega_{ag}t'} \tag{2}$$

where

$$\underline{D}_{ag} = \langle a| \sum_j \underline{r}_j(o) |g\rangle \tag{3}$$

and $\hbar\omega_{ag}$ is the energy difference between state 'a' of the atom and the ground state. It is easy to see that the total probability of transition, when summed over the possible final states $|f\rangle$, is

$$\begin{aligned} \frac{e^2}{\hbar^2} \int_0^t dt' \int_0^t dt'' \, e^{i\omega_{ag}(t'-t'')} \left[D_{ag}\right]_k \left[D_{ag}\right]_\ell^* \Big[&\langle i| E_\ell^{(-)}(\underline{x},t'') E_k^{(+)}(\underline{x},t') |i\rangle \\ &+ \langle i| E_\ell^{(+)}(\underline{x},t'') E_k^{(+)}(\underline{x},t') |i\rangle \\ &+ \langle i| E_\ell^{(-)}(\underline{x},t'') E_k^{(-)}(\underline{x},t') |i\rangle \\ &+ \langle i| E_\ell^{(+)}(\underline{x},t'') E_k^{(-)}(\underline{x},t') |i\rangle \Big] \end{aligned} \tag{4}$$

In the rotating wave approximation (RWA) the term that contributes is

$$\frac{e^2}{\hbar^2 m^2} \int_0^t dt' \int_0^t dt'' \, e^{i\omega_{ag}(t'-t'')} \left[D_{ag}\right]_k \left[D_{ag}\right]_\ell^* \langle i| \, E_\ell^{(-)}(\underline{x},t'') \, E_k^{(+)}(\underline{x},t') \, |i\rangle \tag{5}$$

irrespective of the state $|i\rangle$. When however $|i\rangle$ is restricted to be a pure number state, the transition probability rigorously (ie <u>without</u> RWA) is

$$\frac{e^2}{\hbar^2 m^2} \int_0^t dt' \int_0^t dt'' \, e^{i\omega_{ag}(t'-t'')} \left[D_{ag}\right]_k \left[D_{ag}\right]_\ell^* \Big[\langle i| \, E_\ell^{(-)}(\underline{x},t'') \, E_k^{(+)}(\underline{x},t') \, |i\rangle \tag{6}$$

$$+ \langle i| \, E_\ell^{(+)}(\underline{x},t'') \, E_k^{(-)}(\underline{x},t') \, |i\rangle\Big]$$

The last matrix element can be rewritten as

$$\langle i| \, E_\ell^{(+)}(\underline{x},t'') \, E_k^{(-)}(\underline{x},t') \, |i\rangle = \langle i| \, E_k^{(-)}(\underline{x},t') \, E_\ell^{(+)}(\underline{x},t'') \, |i\rangle + \left[E_\ell^{(+)}(\underline{x},t''), \, E_k^{(-)}(\underline{x},t')\right] \tag{7}$$

The commutator is a c-number independent of $|i\rangle$ and so of no great physical interest. So the matrix elements determining the photodetection are rigorously of the form

$$\langle i| \, E_k^{(-)}(\underline{x},t') \, E_\ell^{(+)}(\underline{x},t'') \, |i\rangle$$

for $|i\rangle$ a pure number state. (It should be clear that the result of equation (6) does not hold for states such as coherent states which have a definite phase.) When $|i\rangle$ represents a <u>single</u> photon we can write such matrix elements as

$$\langle i| \, E_k^{(-)}(\underline{x},t') \, |o\rangle\langle o| \, E_\ell^{(+)}(\underline{x},t'') \, |i\rangle$$

The matrix element $<o|\ E_{\ell}^{(+)}(\underline{x},t")\ |i>$ can be regarded as a photodetection wavefunction. This interpretation becomes clearer if we sum over the states $|a>$ with a weighting function R(a) incorporating the efficiency of detection and work with a single linear polarisation. This gives the probability of detecting a photon to be

$$\int_0^t dt' \int_0^t dt''\ S(t''-t') <i|\ E^{(-)}(\underline{x},t'')\ E^{(+)}(\underline{x},t')\ |i> \tag{8}$$

where

$$S(t) = \int_{-\infty}^{\infty} \frac{e^2}{\hbar^2 m^2} \sum_a R(a)\ \left[D_{ag}\right] \left[D_{ag}\right]^* e^{i\omega_{ag} t} \tag{9}$$

If S(t"−t') = S δ(t"−t') (ie it is broadband), then the photodetection probability simplifies to

$$\int_0^t dt'\ S <i|\ E^{(-)}(\underline{x},t')\ E^{(+)}(\underline{x},t')\ |i> \tag{10}$$

For a one−photon state the probability of detection between t and t+dt is

$$S <i|\ E^{(-)}(\underline{x},t)\ |o><o|\ E^{(+)}(\underline{x},t)\ |i> = S\ \left|<o|\ E^{(+)}(x)\ |i>\right|^2 \tag{11}$$

This gives the photodetection analogue of the traditional concept of a wavefunction which requires the existence of a function $\psi(\underline{x},t)$ such that the probability is $|\psi(\underline{x},t)|^2$. We have so far not specified $|i>$ in detail. We can of course calculate the probability of detecting a photon in any single photon state, but for an arbitrary state this will have no intrinsic interest. It is necessary to have a prescription for calculating the most isotropically localised single photon $|i>$, eg a momentum eigenstate is very unsuitable. The means to construct such a state was pioneered by Jauch and Piron (1967). It goes back to an 'unconventional' formulation of Mackey (1978) (who built on earlier work by Weyl). Mackey introduced the concept of imprimitivity into physics. The photon requires a more general setting which naturally is described as a generalised imprimitivity.

An ordinary system of imprimitivity is a set of quantum-mechanical operators P_S (for all the Borel sets S of the three dimensional (Euclidean) configuration space). $S \to P_S$ is a projective measure and is covariant under the action of the Euclidean group, ie we have

$$P_{\varphi} = 0 \quad , \quad \varphi \text{ is the empty set}$$

$$P_{\mathbb{R}^3} = 1$$

$$P_{S \cap T} = P_S P_T$$

and

$$P_{\bigcup_i S_i} = \sum_i P_{S_i} \quad \text{when} \quad S_i \cap S_j = \varphi \quad \text{for} \quad i \neq j \ .$$

The covariance properties are summarised by

$$U_{\alpha} P_S U_{\alpha}^{-1} = P_{S\alpha^{-1}}$$

where U_{α} is an irreducible representation of any element α in the three dimensional Euclidean group and $S\alpha^{-1}$ is the set of points whose image under α gives S.

A generalised imprimitivity is like an ordinary imprimitivity except that for a sequence of disjoint sets E_i

$$R\left[\sum_i P_{E_i}\right] \subseteq \left[P_{\bigcup E_i}\right] \ .$$

R(.) is the range of the operator which is its argument. For an ordinary imprimitivity the above relation would be strict equality always. A general one-photon state which is as isotropically localised within a Borel set S as possible is given by (Amrein, 1969)

$$\int \frac{d^3p}{(2\pi)^3 \, 2p^o} \left[\varphi_1(p) \, a_1^+(p) + \varphi_{-1}(p) \, a_{-1}^+(p)\right] |o\rangle \qquad (12)$$

where

$$\left[a_{\lambda'}(p'),\ a^{+}_{\lambda}(p)\right] = (2\pi)^3\ 2p^o\ \delta_{\lambda\lambda'}\ \delta^{(3)}(\underline{p}-\underline{p}') \tag{13}$$

$\lambda,\lambda' = \pm 1$ are helicities and p is the 4-vector $(p^o,\underline{p})$. The $\varphi_{\pm 1}$ satisfy the conditions

$$\varphi_{\pm 1}(p) = (\underline{p}^2)^{\frac{1}{4}} \sum_{\beta=-1}^{1} D^1_{\pm 1\beta}(X)\ F^{-1}\ P_S\ F\ S_\beta(p) \tag{14}$$

with

$$S_\beta(p) = \frac{1}{\sqrt{|\underline{p}|}} \left[D^1_{\beta 1}(X^{-1})\ \varphi_1(p) + D^1_{\beta,-1}(X^{-1})\ \varphi_{-1}(p)\right] \tag{15}$$

F is the Fourier transform operator from p to x space, and D^1 are the matrices familiar from angular momentum theory for angular momenta of magnitude one. X is the rotation taking p to $(p^o,o,o,|\underline{p}|)$. $S_\beta(p)$ has three components while a photon has only two helicity degrees of freedom and so it is trying to make the photon look like a massive particle. φ_1 and φ_{-1} are then regarded as the +1 and −1 helicity components of the imagined massive particle in a frame where this particle's momentum is $(p^o,o,o,|\underline{p}|)$. The rotation X^{-1} in $S_\beta(p)$ transforms this artificial massive state into the natural frame where the particle momentum is p. For a massive looking state we can apply an ordinary imprimitivity, viz $F^{-1}\ P_S\ F$. After this operation the state seems to have three spin degrees of freedom which we correct for by the final operations of $D^1_{\pm 1\beta}(X)$. Such states are sometimes called <u>weakly localised</u>. It is possible to calculate the photodetection wavefunction for such a state. Such a calculation (Pike and Sarkar, 1987) shows that the above optimally localised (and hence most particle-like) one-photon state has a spatial dependence of the photodetection probability which falls off as a seventh power of distance, the same as that calculated by Amrein for the energy density.

The pioneering experiment of Burnham and Weinberg (1970) on parametric down-conversion in a nonlinear crystal has recently been revived (Friberg, Hong and Mandel, 1985; Jakeman and Walker, 1985). For usual laser (or pump) powers, since the nonlinearity is very weak, the correlated two-photon states (an idler and a signal photon) that are produced are temporarily well separated on average from other pairs.

Consequently a destructive detection of one photon of the pair leaves a single photon in the field to be detected. Such an experiment provides an opportunity to study the properties of a single photon. A priori such a photon is unlikely to be as localised as one constructed mathematically from a generalised imprimitivity. In the mathematical construction we use as many wavevectors as possible (consistent with the transverse nature of the photon polarisation). The experimental dimensions are such that the correlated photons are each constructed from a certain momentum range around well-defined wavevectors. We ask the question as to how localised the photons can be, given this restriction. Before we can answer this we need to write down a Hamiltonian which describes the quantum-mechanical interaction of light with the crystal. We will take the most obvious and microscopic Hamiltonian H

$$H = \int d^3x \left[\tfrac{1}{2} \left[\frac{\partial}{\partial t} A_k(\underline{x},t) \right]^2 - \tfrac{1}{2} \left[\frac{\partial}{\partial x_i} A_k(\underline{x},t) - \frac{\partial}{\partial x_k} A_i(\underline{x},t) \right]^2 + J_k(\underline{x},t)\, A_k(\underline{x},t) \right] \qquad (16)$$

$$+ H_{crystal}$$

where $H_{crystal}$ is a complicated Hamiltonian describing how the crystal is held together. $\underline{A}(\underline{x},t)$ is the electromagnetic vector potential operator and $\underline{J}$ is the current of (conduction) electrons in the crystal. (It may be possible, as has been tried by some authors, to follow another tack whereby the classical matter–field interaction given by $\underline{P}\cdot\underline{E}$ (where $\underline{P}$, the polarisation, is expressed as a power series in $\underline{E}$) is quantised. However there seem to be some fundamental difficulties (Hillery, and Mlodinow, 1984) associated with this approach. The solution of the Heisenberg equations of motion associated with (16) is

$$A_k(\underline{x},t) = A_k^{(o)}(\underline{x},t) + \iint dt'\, d^3x'\, G^{k\ell}_{(R)}(\underline{x}-\underline{x}',t-t')\, J_\ell(\underline{x}',t') \qquad (17)$$

where

$$G^{k\ell}_{(R)}(\underline{x},t) = -\frac{1}{(2\pi)^4}\int e^{-ip\cdot x}\frac{\left[\delta_{k\ell}-\frac{p_k p_\ell}{\underline{p}^2}\right]}{\left[p_o+i\epsilon\right]^2-\underline{p}^2}\,d^4p \qquad (18)$$

and $A_k^{(o)}(\underline{x},t)$ is the solution of the homogeneous Heisenberg equations (ie for $\underline{J} = 0$). The coincidence rates for the detection of two photons at distinct space–time points is determined by

$$G^{(2,2)}(x_1',x_2';x_2,x_1) = \langle A^{(-)}(x_1')\,A^{(-)}(x_2')\,A^{(+)}(x_2)\,A^{(+)}(x_1)\rangle \qquad (19)$$

Here $x_i = (\underline{x}_i,t_i)$ and $x_i' = (\underline{x}_i',t_i')$, $i = 1,2$.

(The arguments leading to this are very similar to those for single–photon detection and the vector potential rather than the electromagnetic field appears since the interaction Hamiltonian is of the full form $\underline{A}\cdot\underline{J}$ rather than $\underline{E}\cdot\underline{r}$.) If the quantum state of the electromagnetic field has the form

$$|\psi(t)\rangle = \alpha(t)\,|0,0\rangle + \beta(t)\,|1,1\rangle \qquad (20)$$

where $|0,0\rangle$ is a state with no idler or signal photon and $|1,1\rangle$ has one idler and one signal photon, then Mollow (1973) noticed that

$$G^{(2,2)}(x_1',x_2';x_1,x_2) \approx G^{(2,0)}(x_1',x_2')\,G^{(0,2)}(x_2,x_1) \qquad (21)$$

provided $|\beta| << \alpha \sim 1$. The Green functions $G^{(2,0)}(x_1',x_2')$ and $G^{(0,2)}(x_2,x_1)$ are given by

$$G^{(0,2)}(x_2,x_1) = \langle A^{+}(x_2)\,A^{(+)}(x_1)\rangle \qquad (22)$$

and

$$G^{(2,0)}(x_2,x_1) = \langle A^{(-)}(x_2)\,A^{(-)}(x_1)\rangle \qquad (23)$$

where we have suppressed indices. Equation (20) is physically a reasonable approximation for normal pump powers. We can deduce (21) as follows:

$$G^{(2,2)}(x_1',x_2';x_2,x_1)$$

$$= \sum_{n,m} \langle\psi(t)|\ A^{(-)}(x_1')\ A^{(-)}(x_2')\ |n,m\rangle\langle n,m|\ A^{(+)}(x_2)\ A^{(+)}(x_1)\ |\psi(t)\rangle$$

$$= \langle\psi(t)|\ A^{(-)}(x_1')\ A^{(-)}(x_2')\ |0,0\rangle\langle 0,0|\ A^{(+)}(x_2)\ A^{(+)}(x_1)\ |\psi(t)\rangle \qquad (24)$$

$$\simeq \langle\psi(t)|\ A^{(-)}(x_1')\ A^{(-)}(x_2')\ |\psi(t)\rangle\langle\psi(t)|\ A^{(+)}(x_2)\ A^{(+)}(x_1)\ |\psi(t)\rangle$$

The evaluation of $G^{(0,2)}$ is easier than $G^{(2,2)}$ and so (21) is useful. Mollow (1973) produced an elegant analysis of $G^{(0,2)}$ and showed that sufficiently far from the crystal $G^{(0,2)}$ is determined by the positive-frequency part of the time-ordered current correlation function

$$G^{(0,2)}(x_2,x_1) = \iint d^4\bar{x}_2\ d^4\bar{x}_1\ G_R\left[x_2-\bar{x}_2\right]\ G_R\left[x_1-\bar{x}_2\right]\ \cdot\ \langle T\left[J(\bar{x}_2)\ J(\bar{x}_1)\right]\rangle^{++} \qquad (25)$$

The subsequent analysis of this expression by Mollow did not indicate how it is possible to disentangle the effects of linear and nonlinear susceptibility from $\langle T(J(x_2)J(x_1))\rangle$. The linear susceptibility describes the usual complex refractive index of the crystal. We can explicitly remove the linear susceptibility contributions from the current correlation function and obtain a renormalised G_R (Pike and Sarkar, 1988). For large $|\underline{x}|$

$$G(\underline{x},t) \sim \left[\delta_{ij} - \frac{x_i x_j}{|\underline{x}|^2}\right] \frac{1}{|\underline{x}|}\ \delta\left[|\underline{x}| - \frac{c}{n}\ t\right] \qquad (26)$$

where n is the refractive index and c is the speed of light in vaccum. In current experiments the typical dimensions of a crystal are much larger than the mean wavelength of the input pump laser and also the dimensions of the crystal are much smaller than the distance from the crystal to the detectors. Although physically the crystal has sharp edges it is useful to approximate it by a mass density which behaves as

$$\exp\left[-\frac{\left[x_1^2 + x_2^2\right]}{\xi_\perp^2}\right] \exp\left[-\frac{x_3^2}{\xi_3^2}\right] \tag{27}$$

where ξ_3 and $\xi_\perp$ are typical dimensions in the propagation direction of the pump field and in the perpendicular direction. This together with (26) and (25) leads to

$$G_{i\ell}^{(0,2)}(x_2, x_1)$$

$$\sim \int_0^{\omega_o}\int_0^{\omega_o} d\omega_1\, d\omega_2 \int d^3r \exp\left[-\frac{r_1^2+r_2^2}{\xi_\perp^2} - \frac{r_3^2}{\xi_3^2}\right] \exp\left(-i\omega_2(t_2-|\underline{x}_2|) - i\omega_1(t_1-|\underline{x}_1|)\right)$$

$$\exp\left[ir_3\left[k_o - n_2\omega_2 \frac{[\underline{x}_2]_3}{|\underline{x}_2|} - n_1\omega_1 \frac{[\underline{x}_1]_3}{|\underline{x}_1|}\right]\right] \tag{28}$$

$$\exp\left[-ir_1\left[n_2\omega_2 \frac{[\underline{x}_2]_1}{|\underline{x}_2|} + n_1\omega_1 \frac{[\underline{x}_1]_1}{|\underline{x}_1|}\right]\right] F^{jmn}(\omega_1, \omega_2, \underline{x}_1, \underline{x}_2)$$

where

$$F^{jmn}(\omega_1, \omega_2, \underline{x}_1, \underline{x}_2)$$

$$= \omega_1\omega_2\, \delta(\omega_o-\omega_1-\omega_2)\, \chi^{(2)jmn}(\omega_o, \omega_1, \omega_2)\, E\, \frac{1}{|\underline{x}_1||\underline{x}_2|}\left[\delta_{ij} - \frac{[x_2]_i [x_2]_j}{|\underline{x}_2|^2}\right]$$

and the quantum state of the incoming pump is given by a coherent state with amplitude

$$E \exp(ik_o r_3) \exp(i\omega_o t)$$

(The field could have a transverse profile but this is incorporated through (27).) $\chi^{(2)}$ is the nonlinear susceptibility tensor. On performing the r integration we obtain as a factor in the remaining ω integration (Pike and Sarkar, 1987)

$$\exp\left[-\frac{\xi_3^2}{4}\left[k_o - \frac{n_2\omega_2}{c}\frac{[x_2]_3}{|\underline{x}_2|} - \frac{n_1\omega_1}{c}\frac{[x_1]_3}{|\underline{x}_1|}\right]^2\right]$$

$$\cdot \exp\left[-\frac{\xi_\perp^2}{4}\left[\frac{n_2\omega_2}{c}\frac{[x_2]_1}{|\underline{x}_2|} + \frac{n_1\omega_1}{c}\frac{[x_1]_1}{|\underline{x}_1|}\right]^2\right]$$

Since ξ_3 and $\xi_\perp$ are large compared to k_o^{-1}, the dominant contribution to the ω integration is given by

$$k_o = \frac{n_2\omega_2}{c}\cos\theta_2 + \frac{n_1\omega_1}{c}\cos\theta_1 \tag{29}$$

$$n_2\omega_2\sin\theta_2 = n_1\omega_1\sin\theta_1 \tag{30}$$

where

$$\frac{[\underline{x}_i]_3}{|\underline{x}_i|} = \cos\theta_i \quad , \quad i = 1,2$$

(31)

$$\frac{[\underline{x}_i]_1}{|\underline{x}_i|} = (-1)^i\sin\theta_i \quad , \quad i = 1,2$$

The equations (29) and (30) are the phase-matching conditions. They would be strictly valid in the limit $\xi_3 k_o$, $\xi_\perp k_o \to \infty$. Given any two values of ω_1 and ω_2, ω_1^* and ω_2^* say (satisfying $\omega_1^*+\omega_2^* = \omega_o$), it is possible to find θ_1^* and θ_2^* satisfying (29) and (30). Mollow (1973) considered this nearly phase-matched situation and approximated (28) by

$$G_{i\ell}^{(0,2)}(x_2, x_1)$$

$$\sim \int_0^{\omega_o}\int_0^{\omega_o} d\omega_1\, d\omega_2 \int d^3r \exp\left[-\frac{r_1^2+r_2^2}{\xi_\perp^2} - \frac{r_3^2}{\xi_3^2}\right]$$

$$\exp\left(-i\omega_2(t_2-|\underline{x}_2|) - i\omega_1(t_1-|\underline{x}_1|)\right) \qquad (32)$$

$$\exp\left(ir_3(k_o - n_2\omega_2\cos\theta_2^* - n_1\omega_1\cos\theta_1^*)\right)$$

$$\exp\left(-ir_1(n_2\omega_2\sin\theta_2^* - n_1\omega_1\sin\theta_1^*)\right)$$

$$\omega_1^*\omega_2^*\ \delta(\omega_o-\omega_1-\omega_2)\ \chi^{(2)jmn}(\omega_o,\omega_1^*,\omega_2^*)\ E\ \frac{1}{|\underline{x}_1||\underline{x}_2|}\left[\delta_{ij} - \frac{[x_2]_i[x_2]_j}{|\underline{x}_2|^2}\right]$$

(where the two detectors are taken to be in the plane $(x_1)_2 = (x_2)_2 = 0$). The full ω dependence is kept only in the exponent since in the phase-matched situation this is the most rapidly varying term. Since we assume that the frequencies that we are concerned with are far away from any crystal resonances, it is satisfactory to consider $\chi^{(2)jmn}$ to be independent of frequency. However in the search for photon tails it is <u>not</u> adequate as done by Mollow to replace the $\omega_1\omega_2$ in (28) by $\omega_1^*\omega_2^*$. In fact in this example the interesting aspect of the photon tail comes precisely from the $\omega_1\omega_2$ factor. After performing the r integration the ω_1 and ω_2 integration is

$$\int_0^{\omega_o}\int_0^{\omega_o} d\omega_1\, d\omega_2 \exp\left[-\frac{\xi_3^2}{4}\left[k_o - \frac{n_2\omega_2}{c}\frac{(x_2)_3}{|\underline{x}_2|} - \frac{n_1\omega_1}{c}\frac{(x_1)_3}{|\underline{x}_1|}\right]^2\right]$$

$$\exp\left[-\frac{\xi_\perp^2}{4}\left[\frac{n_2\omega_2}{c}\frac{(x_2)_1}{|\underline{x}_2|} + \frac{n_1\omega_1}{c}\frac{(x_1)_1}{|\underline{x}_1|}\right]^2\right] \tag{33}$$

$$\omega_1\omega_2 \exp\left[-i\omega_2\tilde{t}_2\right] \exp\left[-i\omega_1\tilde{t}_1\right] \delta(\omega_o-\omega_1-\omega_2)$$

with

$$\tilde{t}_i = t_i - n_i\frac{|\underline{x}_i|}{c} , \quad (i = 1,2) .$$

The ω_1 integration is automatic with the use of the δ-function. The Fourier theorem concerning convolutions allows as to the rewrite the ω_2 integration as

$$\exp\left[-i\omega_o\tilde{t}_1\right] \int_{-\infty}^{\infty} d\tau\; I_1\left[\tilde{t}_2-\tilde{t}_1-\tau\right] I_2(\tau) \tag{34}$$

where

$$I_1(\tau) = \int_{-\infty}^{\infty} d\omega\; \theta(\omega)\; \theta(\omega_o-\omega)\; (\omega_o-\omega)\omega\; \exp(-i\omega\tau) \tag{35}$$

and

$$I_2(\tau) = \int_{-\infty}^{\infty} d\omega \exp\left[-\frac{1}{4}\left[U\omega^2-2V\omega\right] - i\omega\tau\right] \tag{36}$$

Here

$$U = \xi_3^2 \left[n_2\cos\theta_2^* - n_1\cos\theta_1^*\right]^2 + \xi_\perp^2 \left[n_2\sin\theta_2^* + n_1\sin\theta_1^*\right]^2 \tag{37}$$

and

$$V = \xi_3^2 \left[k_o - n_1\omega_o\cos\theta_1^*\right] \left[n_2\cos\theta_2^* - n_1\cos\theta_1^*\right]$$

$$+ \xi_\perp^2 n_1\omega_o\sin\theta_1^* \left[n_2\sin\theta_2^* + n_1\sin\theta_1^*\right] \tag{38}$$

It is easy to show that

$$I_1(\tau) = - \frac{2\omega_o}{\tau^2} e^{-i\tau\frac{\omega_o}{2}} \left[\cos\frac{\tau\omega_o}{2} - \frac{\sin\frac{\tau\omega_o}{2}}{\frac{\tau\omega_o}{2}}\right] \tag{39}$$

and

$$I_2(\tau) = \sqrt{\frac{4\pi}{U}} \exp\left[\frac{1}{4}\frac{V^2}{U}\right] \exp\left[-\frac{\tau^2}{U}\right] \exp\left[-i\frac{V\tau}{U}\right] \tag{40}$$

Clearly I_2 is exponentially localised. I_1 provides an oscillating power-law tail. In contrast to the approximately localised photon formed through a generalised imprimitivity which has a photodetection probability falling off as τ^{-7}, the asymptotic behaviour in parametric down-conversion is

$$\frac{1}{\tau^4}\left[\cos\frac{\tau\omega_o}{2} - \frac{\sin\frac{\tau\omega_o}{2}}{\frac{\tau\omega_o}{2}}\right]^2 \tag{41}$$

Moreover this localisation is highly anisotropic. In directions perpendicular to those defined by θ_1^* and θ_2^* there is, in lowest order, no fall-off. This is because the approximate phase matching allows only a limited directionality of wavevectors to contribute to the idler and signal photons. In the directions θ_1^* or θ_2^* although a range of frequencies are allowed, the range is not large enough to obtain the weakly localised states of Jauch, Piron and Amrein. The main reason for this restricted range is due to the details of the parametric down–conversion experiment.

We are thus still some way off from producing the most localised photons possible in the laboratory. Indeed it would seem that any experiments, such as parametric down–conversion, which have a preferred direction of wavevector (such as in the input pump) are not likely to have available through atom–field interactions enough different types of wavevector to give a localised photon states approaching the optimal weakly localised states.

ACKNOWLEDGEMENTS

We would like to thank C. M. Caves, J. G. Rarity and P. R. Tapster for discussions.

REFERENCES

Amrein, W.O. 1969, Helv. Phys. Acta 42 149.
Burnham, D.C. and Weinberg, D.L. 1970, Phys. Rev. Lett. 25 84.
Friberg, S., Hong, C.K. and Mandel, L. 1985, Phys. Rev. Lett. 54 2011.
Glauber, R.J. 1963, Phys. Rev. 131 2766.
Hillery, M. and Mlodinow, L.D. 1984, Phys. Rev. A30 1860.
Jakeman, E. and Walker, J.G. 1985, Opt. Commun. 55 219.
Jauch, J.M. and Piron, C. 1967, Helv. Phys. Acta 40 559.
Mackey, G.W. 1978, Unitary Group Representations in Physics, Cumming, Reading.
Mollow, B.R. 1973, Phys. Rev. A8 2684.
Newton, T.D. and Wigner, E.P. 1949, Rev. Mod. Phys. 21 400.
Pike, E.R. 1986, Coherence, Co–operation and Fluctuations, Ed. H. Haake, L.M. Narducci and D.F. Walls, Cambridge University Press, Cambridge.
Pike, E.R. and Sarkar, S. 1986, Frontiers in Quantum Optics, Ed. E.R. Pike and S. Sarkar, Hilger, Bristol.
Pike, E.R. and Sarkar, S. 1987, Phys. Rev. A35 926.
Pike, E.R. and Sarkar, S. 1987, Power Law Tails of Single Photon States in Parametric Down Conversion, RSRE Malvern Preprint.
Pike, E.R. and Sarkar, S. 1988, Photon Localisation in Parametric Down Conversion, RSRE Malvern Preprint
Rarity, J.G., Tapster, P.R. and Jakeman, E. 1987, Opt. Commun. 62 201.
Wightman, A.S. 1962, Rev. Mod. Phys. 34 845.

QUANTUM NOISE REDUCTION ON TWIN LASER BEAMS

E GIACOBINO, C FABRE, S REYNAUD, A HEIDMANN AND R HOROWICZ

Laboratoire de Spectroscopie Hertzienne de l'E.N.S.
Université Pierre et Marie Curie, T12-E01,
4, place Jussieu 75252 Paris Cedex 05, France.

1. INTRODUCTION

Among the processes which generate light with "non classical" properties, the optical parametric interaction turns out to be a prototypic one (Louisell et al. 1961, Takahasi 1965, Mollow et al. 1967, Mollow 1973, Stoler 1974, Graham 1973). The process involves generation of two "signal" fields at frequencies ω_1 and ω_2 by a non-linear crystal irradiated with a pump field at frequency ω_0 (energy conservation requires that $\omega_0 = \omega_1 + \omega_2$). The crystal can be placed in a resonant optical cavity operating either below or above oscillation threshold depending on the pump power.

Up to now, most of the work have concentrated on Quasi Degenerate Optical Parametric Amplifiers (OPA) (below threshold operation with $\omega_1 \approx \omega_2$), which have been shown both theoretically (Milburn et al. 1981, Lugiato et al. 1982, Yurke 1984, Collett et al. 1984, Collett et al. 1985) and experimentally (Wu et al. 1986) to yield squeezed states of light. Actually the non-linear interaction taking place in the

OPA leads to deamplification of the vacuum fluctuations on some quadrature components of the electromagnetic fields going out of the cavity.

It is the purpose of this paper to give a review of both theoretical and experimental results that we have obtained on the Non-Degenerate Optical Parametric Oscillator (OPO) (above threshold operation with $\omega_1 \neq \omega_2$).In particular it will be shown that most of its characteristics can be understood from the fact that the non-linear crystal emits highly correlated twin photons. The correlation properties between the intensities of the corresponding twin fields are the basic features studied in this paper.

In the theoretical section, we will present three different approaches. The first one is a simple treatment in terms of "corpuscular" photons which provides a good physical insight into the problem (§ 2). The second one is intended to give a proof of the corpuscular model in a fully quantum manner (§ 3). The third one is based upon a classical representation of the fields but in which vacuum fluctuations are properly accounted for (§ 4). This "classical" approach yields the noise spectra of the various measurable quantities in the general case. In the fifth section, we will discuss the experiment and present our observation of quantum noise reduction on the difference between the intensities of the two twin beams (§ 5). The last section will deal with new possible applications of the twin photons to spectroscopy and to the generation of amplitude squeezed laser beams (§ 6).

2. CORPUSCULAR MODEL

We first present a "corpuscular" model(Reynaud 1987) in which photons are considered as classical particles. We consider only the case of non-degenerate emission in which the two types of signal photons can be distinguished by either theirfrequencies ($\omega_1 \neq \omega_2$)or their polarizations. It has been shown both experimentally (Burnham et al. 1970, Friberg et al 1985) and theoretically (Mollow et al. 1967, Mollow 1973, Graham 1984, Hong et al. 1985) that in parametric down conversion the non-linear crystal emits pairs of simultaneous photons 1 et 2 each time a pump photon is annihilated. Such a twin photon generator can be used as the pumping mechanism for an optical cavity resonant at the signal frequencies ω_1 and ω_2. Above some pump power threshold, the system oscillates and yields "twin photon" beams B_1 and B_2 i.e. two laser-like beams having highly correlated intensities. Actually, since the twin-photons produced in the crystal do not necessarily go out of the cavity at the same time, the cavity induces some decorrelation between the twin beams : the numbers of photons delivered in the two output beams are expected to be nearly equal only when counted during a time long compared to the cavity storage time. In other words, measuring the noise spectrum $S_I(\omega)$ of the difference I between the signal beam intensities must reveal a reduction of photon noise inside the cavity bandwith.

The purpose of this section is to give a quantitative prediction for the noise spectrum $S_I(\omega)$. In the present model we will assume that the emissions of the various pairs of photons are randomly distributed. The mean rate of pair emission is denoted A. In addition each collision of any signal photon with the coupling mirror is described by a probability R for being reflected, a probability T for being transmitted (with R + T = 1).

In this framework, we can evaluate the spectrum $S_I(\omega)$ corresponding to the detection scheme sketched in fig. 1 :

$$S_I(\omega) = \frac{1}{\Delta t}\langle|\tilde{I}(\omega)|^2\rangle \tag{1}$$

$$\text{with } \tilde{I}(\omega) = \int_{\Delta t} I(t)\ \exp(-i\omega t)\ dt \tag{2}$$

$$\text{and } I(t) = I_1(t) - I_2(t) \tag{3}$$

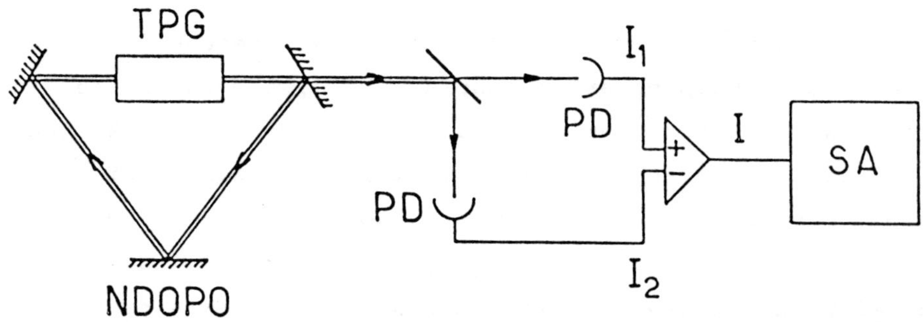

Figure 1: The non-degenerate optical parametric oscillator (NDOPO) is constituted by a twin-photon generator (TPG) inserted in an optical cavity. The twin signal beams are separated and their intensities I_1 and I_2 are measured by photodetectors (PD). The spectrum analyzer (SA) gives the noise spectrum $S_I(\omega)$ of the intensity difference $I = I_1 - I_2$.

In these expressions, $I_1(t)$ and $I_2(t)$ are the instantaneous signal beam intensities monitored by the two photodetectors ; $\tilde{I}(\omega)$ is the Fourier transform of I(t) ; the integration time Δt is much longer than any other characteristic time ; the symbol $\langle\ \rangle$ in eq.(1) describes a "classical" mean value taken over the various possible events.

A particular event can be described by the following random variables : the emission time of a pair, the numbers k_1 and k_2 of reflections on the coupling mirror of the photons 1 and 2. If the intensities are measured as detection rates (and if thequantum efficiencies are assumed to be 1), the contribution of this particular event to I(t) is :

$$I(t) = \delta(t - (t_0+k_1\tau)) - \delta(t - (t_0+k_2\tau)) \qquad (4)$$

where δ is the Dirac function ; t_0 the first possible detection time (no reflection on the coupling mirror), τ is the roundtrip time of the photons inside the cavity (for simplicity, we suppose that t_0 and τ are the same for the two signal modes). The contribution of this event to the Fourier transform $|\tilde{I}(\omega)|^2$ is thus

$$\begin{aligned} |\tilde{I}(\omega)|^2 &= |\exp(-i\omega(t_0+k_1\tau)) - \exp(-i\omega(t_0+k_2\tau))|^2 \\ &= 2[1-\cos((k_1 - k_2)\omega\tau)] \end{aligned} \qquad (5)$$

We now obtain the spectrum $S_I(\omega)$ (Eq.1) by summing these contributions (Eq.5) weighted by the probability TR^{k_1} TR^{k_2} that the photons 1 and 2 undergo respectively k_1 and k_2 reflections and then multiplying by the number A Δt of pair emissions during the integration time :

$$S_I(\omega) = 2A \sum_{k_1=0}^{\infty} \sum_{k_2=0}^{\infty} (TR^{k_1})(TR^{k_2})[1 - \cos(k_1-k_2)\omega\tau] \qquad (6)$$

The summation of the series appearing in Eq.6 leads to :

$$S_I(\omega) = S_0\left(1 - \frac{T^2}{T^2 + 4R\sin^2\left(\frac{\omega\tau}{2}\right)}\right) \qquad (7)$$

with $S_0 = 2A$ (8)

It clearly appears on Eq.7 that photon noise is completely suppressed at $\omega = 0$ ($S_I(0) = 0$) which corresponds to the qualitative discussion given before. More generally, photon noise is suppressed at all the frequencies which are multiples of the free spectral range $\Omega = 2\pi/\tau$. Suppression remains effective inside the Airy peaks (Fig.2). Outside these peaks, $S_I(\omega)$ reproduces the usual photon noise S_0 corresponding to the sum of the intensities of the two beams ($S_0 = \langle I_1 + I_2 \rangle = 2A$).

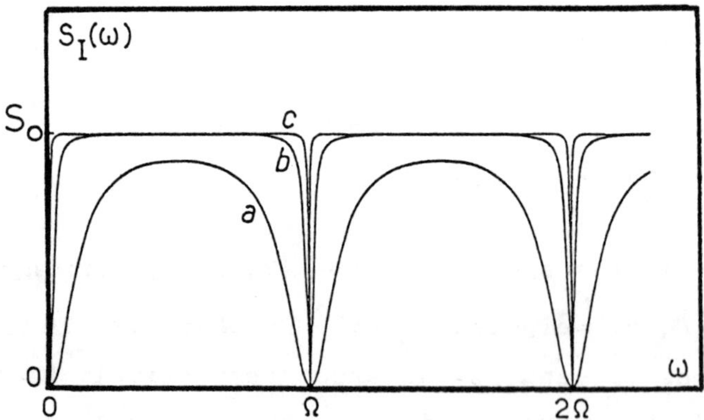

Figure 2 : Predicted noise spectrum $S_I(\omega)$ for various values of the mirror transmission : T = 50% (a), 10% (b), 2% (c). S_0 is the usual photon noise. The frequency unit Ω is the cavity free spectral range.

For a high finesse cavity ($T \ll 1$), the Airy peaks are

well approximated by Lorentzian peaks ; one gets for $\omega \ll 1/\tau$:

$$S_1(\omega) = S_0 \frac{(\omega\tau/T)^2}{1+(\omega\tau/T)^2} \tag{9}$$

The width of the Lorentzian dip is thus directly related to the cavity storage time τ/T, as expected from the previous qualitative discussion.

3. QUANTUM DERIVATION OF THE NOISE SPECTRUM

We now show that the expressions obtained in the corpuscular model can be demonstrated in a fully quantum manner (Reynaud 1987).

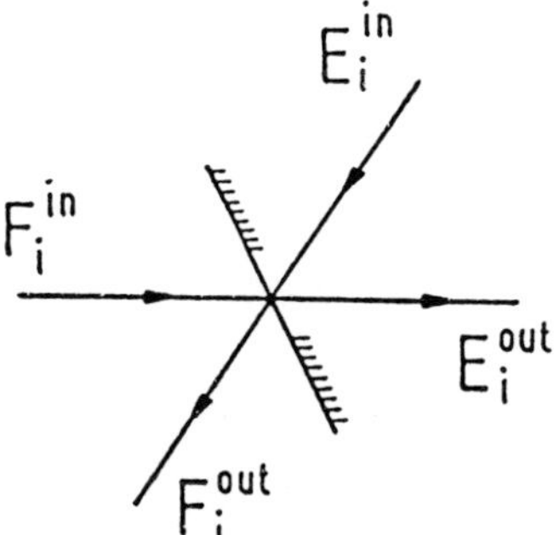

Figure 3 : Reflection-transmission on the coupling mirror as a scattering process. E_i *and* F_i *represent the signal fields (i = 1,2) respectively outside and inside the cavity.*

First, we treat reflection-transmission on the coupling mirror as a scattering problem. The quantum scattering process is described as in classical electromagnetism by transmission and reflection amplitudes $\sqrt{T}$

and $\sqrt{R}$ (phase shifts play no role in the following and they are ignored) :

$$F_i^{out} = \sqrt{R}\, F_i^{in} + \sqrt{T}\, E_i^{in}$$

$$E_i^{out} = \sqrt{T}\, F_i^{in} - \sqrt{R}\, E_i^{in} \qquad (10)$$

In these equations, E and F denote the quantum fields (positive frequency parts of the electric field operators) ; E_i and F_i correspond respectively to the signal fields (i=1,2) outside and inside the cavity ; the labels "in" and "out" refer to the incoming and outgoing waves (see Fig.3). We will assume that input fields E_i^{in} are in the vacuum state (no photon coming onto the coupling mirror from the outside of the cavity). In this case, one can deduce the following equations for the intensity operators :

$$I_i = E_i^{+} E_i \ ; \ J_i = F_i^{+} F_i$$

$$J_i^{out} = R\, J_i^{in} + K_i$$

$$I_i^{out} = T\, J_i^{in} - K_i \qquad (11)$$

with $K_i = \sqrt{RT}\,(E_i^{in+} F_i + F_i^{in+} E_i)$

The intensity operators I_i^n do not appear ($I_i^n = 0$). The operators K_i are "delta-correlated noise" operators :

$$\langle K_i \rangle = 0$$
$$\langle K_i(t)\, K_j(t') \rangle = D\, \delta_{ij}\, \delta(t-t') \qquad (12)$$

with $D = RT \langle J_1^{in} \rangle$

In other words, the operators K_i are characterized by a white noise spectrum (see for example Gardiner 1983) :

$$\langle \tilde{K}_i(\omega) \, \tilde{K}_j(\omega') \rangle = 2\pi \, D \, \delta_{ij} \, \delta(\omega+\omega') \tag{13}$$

($\tilde{K}_i(\omega)$ is the Fourier transform of $K_i(t)$).

We now come to the quantum treatment of the parametric interaction. We will use only the assumption that the signal photons are emitted by pairs. So the parametric interaction is described by some hamiltonian which commutes with the difference between the two signal intensities (Graham 1984) and we can write :

$$J_1^{in}(t) - J_2^{in}(t) = J_1^{out}(t-\tau) - J_2^{out}(t-\tau) \tag{14}$$

where τ is the roundtrip time for the signal photons.

Iterating Eqs.(14) and (11) leads to :

$$J_1^{in}(t) - J_2^{in}(t) = K(t-\tau) + RK(t-2\tau) + R^2K(t-3\tau) + \ldots \tag{15}$$

with $K(t) = K_1(t) - K_2(t)$

One then gets the operator $I(t)$ corresponding to the detection scheme sketched in Fig.1 :

$$I(t) = I_1^{out}(t) - I_2^{out}(t) = -K(t)+T[K(t-\tau)+RK(t-2\tau)+\ldots] \tag{16}$$

The Fourier transform $\tilde{I}(\omega)$ of I(t) is :

$$\tilde{I}(\omega) = \alpha(\omega)\tilde{K}(\omega)$$

with $\alpha(\omega) = -1 + T[e^{-i\omega\tau}+Re^{-2i\omega\tau}+R^2e^{-3i\omega t}+...]$

$$= -1 + Te^{-i\omega\tau}/(1 - Re^{-i\omega\tau}) \quad (17)$$

The expression of the noise spectrum $S_I(\omega)$ is obtained from :

$$\langle\tilde{I}(\omega)\ \tilde{I}(\omega')\rangle = 2\pi S_I(\omega)\delta(\omega+\omega') \quad (18)$$

Comparing Eq.(18) with Eqs.(13), (15) and (17) gives :

$$S_I(\omega) = 2D|\alpha(\omega)|^2 \quad (19)$$

One easily checks that this expression is identical to the result of the "corpuscular" model (noting that D = RA).

The quantum treatment not only provides a justification of the corpuscular model but it also gives an interesting interpretation of the quantum noise reduction. In this approach, quantum noise is clearly a consequence of the vacuum fluctuations going onto the coupling mirror (fields E_1^{in} ; see Fig.3) which are in fact the only source of fluctuations. Quantum noise reduction appears to be due to a destructive interference between the effect of the vacuum fluctuations reflected on the mirror and those which have passed through the cavity. Noise is totally suppressed at ω=0 where the interference is perfectly destructive ($\alpha(0)=0$, in Eq.(17)).

We have considered a perfect cavity, where loss of photons is only due to escape through the coupling mirror.

Other loss mechanisms will obviously degrade the quantum noise reduction . Intracavity losses, crystal absorption, stray reflections, scattering should ideally be kept much smaller than the transmission of the coupling mirror. Extracavity losses, including those coming from the detectors quantum efficiencies have to be much smaller than 1. The effect of the losses can be measured by a parameter η, probability for any signal photon emitted by the parametric device to be effectively registered by a detector. As it is well known for the effect of an imperfect quantum efficiency, the photon noise reduction factor cannot be better than the "loss per photon" $(1-\eta)$. It would be possible to take these losses rigorously into account in the preceding treatment. We prefer to derive these results with the semiclassical method that we expose in the next paragraph. As a matter of fact, the latter method is well adapted to the prediction of various intensity as well as phase noise spectra in more general situations, for example when the loss parameters are different for the two generated fields.

4. SEMI-CLASSICAL METHOD

The fluctuation characteristics of the OPO emission can be computed very simply using the following "semi-classical" approach (Reynaud et al. 1987). The dynamics of small field fluctuations are described by linearizing the classical equations of motion in the vicinity of the stationary state. We consider that these field fluctuations are driven by the vacuum fluctuations entering the cavity through the coupling mirror. It can be shown that standard quantum methods assuming ideal "noise" (Savage et al. 1987,

Lane et al. 1988, Björk et al. 1988) lead to the same equations. Also, we will see that for the noise spectrum of the intensity difference (without losses), the result is identical to the quantum result (Eq. (19) above).

As before, we consider that the signal fields are confined in an optical ring cavity ; the two fields can be separated from each other by either their frequencies ω_1 and ω_2 (with ω_1 close to ω_2) or their polarizations (which will be the experimental case). The pump field at frequency ω_0 is also confined in the cavity, and we allow the transmission coefficients of the coupling mirror T_i to be different from one another for the three fields (i = 0,1,2). Other internal losses in the cavity will be accounted for. Perfect phase matching is assumed, as well as exact resonance with the cavity modes for the pump and signals. If the one-pass gain and losses are small, the equations of motions for the classical amplitudes α_1, α_2, α_0 associated with the annihilation operators a_1, a_2, a_0 can be written as differential equations :

$$\left\{\begin{aligned} \tau\dot{\alpha}_1 + (\gamma_1 + \mu_1)\alpha_1 &= 2\chi\alpha_0\alpha_2^* + \sqrt{2\gamma_1}\,\alpha_1^{in} + \sqrt{2\mu_1}\,\beta_1^{in} \\ \tau\dot{\alpha}_2 + (\gamma_2 + \mu_2)\alpha_2 &= 2\chi\alpha_0\alpha_1^* + \sqrt{2\gamma_2}\,\alpha_2^{in} + \sqrt{2\mu_2}\,\beta_2^{in} \\ \tau\dot{\alpha}_0 + (\gamma_0 + \mu_0)\alpha_0 &= -2\chi\alpha_1\alpha_2 + \sqrt{2\gamma_0}\,\alpha_0^{in} + \sqrt{2\mu_0}\,\beta_0^{in} \end{aligned}\right. \qquad (20)$$

The cavity round-trip time τ is assumed to be the same for all three fields for the sake of simplicity. The one-pass loss parameters γ_i and μ_i (i = 0,1,2) are associated respectively with the output mirror (the γ_i are related to the mirror transmissions T_i through $T_i = 2\gamma_i$) and with other loss mechanisms (the μ_i are due to crystal absorption, surface scattering and imperfections of the other mirrors). The total loss coefficients $\gamma_i + \mu_i$ will be denoted γ'_i.

The quantities α_i^{in} and β_i^{in} (i=0,1,2) are the incoming fields, entering the cavity respectively through the coupling mirror and by the same mechanism that causes the internal losses. In the following, only the incoming pump field α_o^{in} has a non zero mean value. The other incoming fields will be treated as classical fluctuating fields around a zero mean value with standard deviations equal to those of the quantum vacuum fields.

From the values of the fields α_i inside the cavity, we will then deduce the outgoing fields α_i^{out} by the following equations :

$$\alpha_i^{out} = \sqrt{2\gamma_i}\ \alpha_i - \alpha_i^{in} \qquad (21)$$

4.1 Stationary values

The $\bar{\alpha}_i$'s , stationary solutions for the α_i's can easily be calculated by taking $\alpha_i^{in} = \beta_i^{in} = 0$ (i=1,2). A choice of phases is made, where $\bar{\alpha}_o$, $\bar{\alpha}_1$ and $\bar{\alpha}_2$ are real. Actually, whereas the phase of either $\bar{\alpha}_o$ or α_o^{in} can be arbitrarily chosen, the equations do not fix the value of the phase difference between α_1 and α_2. We will come back to the phase diffusion associated with this feature.

Above threshold, there is a non-trivial solution to Eqs. (20), given by :

$$\gamma'_i \quad \bar{\alpha}_i^2 = \frac{\gamma'_o \; \gamma'_1 \; \gamma'_2}{4 \; \chi^2} (\sigma - 1) \qquad (i=1,2) \tag{22-a}$$

$$\bar{\alpha}_0^{\;2} = \frac{\gamma'_1 \; \gamma'_2}{4 \; \chi^2} \tag{22-b}$$

where :

$$\sigma = \frac{2\sqrt{2}\chi}{\sqrt{\gamma'_o \; \gamma'_1 \; \gamma'_2}} \alpha_o^{in} \tag{23}$$

is larger than 1 above threshold. It can be written as :

$$\sigma = \sqrt{\frac{P}{P_{th}}} \tag{24-a}$$

where : $P_{th} = \dfrac{\gamma'_0 \; \gamma'_1 \; \gamma'_2}{8 \; \chi^2}$ (24-b)

is the pump intensity threshold and P the incident pump intensity.

Let us note that the outgoing signal intensities

$$\bar{I}_1^{out} = 2\gamma_1 \; \bar{\alpha}_1^2 \tag{25}$$

obey the simple relation :

$$\frac{\bar{I}_1^{out}}{\bar{I}_2^{out}} = \frac{\gamma_1 \; \gamma'_2}{\gamma_2 \; \gamma'_1} \tag{26}$$

There are equal in the absence of internal losses ($\mu_1 = 0$; $\gamma'_1 = \gamma'$) even if the mirror transmissions γ_1 and γ_2 are different.

4.2 Equations for the fluctuations

We linearize Eqs. (20) around the mean values given by Eqs. (22). Setting $\alpha_1 = \overline{\alpha}_1 + \delta\alpha_1$, we obtain :

$$\tau\delta\dot{\alpha}_1 + \gamma'_1\,\delta\alpha_1 = \sqrt{\gamma'_1\,\gamma'_2}\;\delta\alpha^*_2 + \sqrt{\gamma'_0\,\gamma'_1\,(\sigma-1)}\;\delta\alpha_0 + \sqrt{2\gamma_1}\;\alpha^{in}_1 + \sqrt{2\mu_1}\;\beta^{in}_1 \tag{27-a}$$

$$\tau\delta\dot{\alpha}_2 + \gamma'_2\,\delta\alpha_2 = \sqrt{\gamma'_2\,\gamma'_1}\;\delta\alpha^*_1 + \sqrt{\gamma'_0\,\gamma'_2\,(\sigma-1)}\;\delta\alpha_0 + \sqrt{2\gamma_2}\;\alpha^{in}_2 + \sqrt{2\mu_2}\;\beta^{in}_2 \tag{27-b}$$

$$\tau\delta\dot{\alpha}_0 + \gamma'_0\,\delta\alpha_0 = -\sqrt{\gamma'_0\,\gamma'_1\,(\sigma-1)}\;\delta\alpha_2 - \sqrt{\gamma'_0\,\gamma'_2\,(\sigma-1)}\;\delta\alpha_1 + \sqrt{2\gamma_0}\;\delta\alpha^{in}_0 + \sqrt{2\mu_0}\;\beta^{in}_0 \tag{27-c}$$

Introducing the real and imaginary parts of the fields, we set :

$$\begin{cases} p_1 = \delta\alpha_1 + \delta\alpha^*_1 \\ q_1 = -i(\delta\alpha_1 - \delta\alpha^*_1) \end{cases} \tag{28}$$

with similar notations for the input and output fields.

As we have chosen the stationary solutions to be real,

p_1 gives the amplitude fluctuations, while q_1 is proportional to the phase fluctuations. The differential equations (27) are then transformed into algebraic equations for the Fourier components $\tilde{p}_1^{out}(\omega)$, $\tilde{q}_1^{out}(\omega)$ as a function of the input fluctuations by using eq.(21). With a proper choice of the normalization factor for the input fluctuations variances :

$$V_{p_1}^{in}(\omega) = \langle|\tilde{p}_1^{in}(\omega)|^2\rangle = 1 \qquad (29)$$

the squeezing $S_{p_1}(\omega)$ will be directly given by the output fluctuation variance :

$$S_{p_1}(\omega) = V_1^{out}(\omega) = \langle|p_1^{out}(\omega)|^2\rangle \qquad (30)$$

4.3 Fluctuation spectra in the balanced case

We will first give the results in the simple case in which the losses are the same for the two signal fields ($\gamma_1 = \gamma_2 = \gamma$, $\mu_1 = \mu_2 = \mu$, $\gamma_1' = \gamma_2' = \gamma'$).

It can immediatly be seen on Eqs. (27-a) and (27-b) that the pump fluctuations will not be coupled into the fluctuations $\delta\alpha_1 - \delta\alpha_2$. Thus for the difference of the amplitude fluctuations :

$$r = \frac{1}{\sqrt{2}}(p_1 - p_2) \qquad (31)$$

(with similar notations for the input and output fields), we get the following equation:

$$\tau\dot{r} + 2\gamma' r = \sqrt{2\gamma}\, r^{in} + \sqrt{2\mu}\, r'^{in} \qquad (32)$$

We thus obtain for the Fourier component $\tilde{r}(\omega)$ of r :

$$\tilde{r}(\omega) = \frac{1}{2\gamma' + i\omega\tau}\left(\sqrt{2\gamma}\,\tilde{r}^{in}(\omega) + \sqrt{2\mu}\,\tilde{r}'^{in}(\omega)\right) \quad (33)$$

The outgoing field r^{out} is easily derived from (10) :

$$r^{out} = \sqrt{2\gamma}\, r - r^{in} \quad (34)$$

Hence combining Eqs (33) and (34) leads to :

$$r^{out}(\omega) = \frac{2\gamma - 2\gamma' - i\omega\tau}{2\gamma' + i\omega\tau}\,\tilde{r}^{out}(\omega) + \frac{2\sqrt{\gamma\mu}}{2\gamma' + i\omega\tau}\,\tilde{r}^{out}(\omega) \quad (35)$$

Since the variances of the input vacuum fluctuations are normalized to 1, the squeezing on the r component is simply given by the variance of

$$S_r(\omega) = V_r^{out}(\omega) = \frac{4\mu\gamma' + \omega^2\tau^2}{4\gamma'^2 + \omega^2\tau^2} \quad (36)$$

In the present case, the mean intensities on the twin beams are equal and the noise spectrum on the intensity difference $S_I(\omega)$ is proportional to $S_r(\omega)$:

$$S_I(\omega) = S_0 \frac{4\mu\gamma' + \omega^2\tau^2}{4\gamma'^2 + \omega^2\tau^2} \quad (37)$$

where S_0 is the shot noise on a beam with a total intensity equal to the sum of the intensities of the twin beams. In the absence of internal losses ($\mu=0, \gamma'=\gamma$), one gets the same result as in the previous section (recalling that $T=2\gamma$). It can be seen that internal losses decrease the amount of noise reduction at $\omega=0$ by a factor :

$$\frac{\mu}{\gamma'} = \frac{\mu}{\mu+\gamma} \tag{38}$$

Note that this factor is just the proportion of emitted photons which do not reach the photodetector (they are for example scattered by the crystal or emitted through a mirror different from the output one). This implies that extra losses associated with μ have to be kept much smaller than losses through the output mirror. The noise spectrum is shown in Fig.4 for various values of μ.

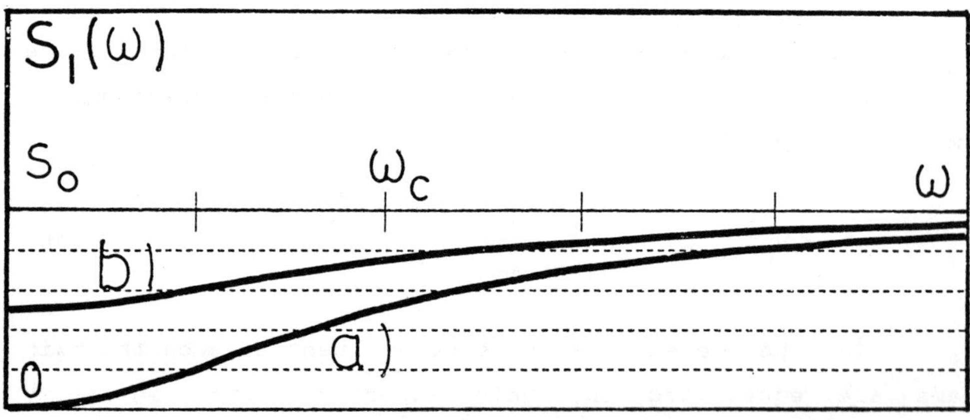

Figure 4 :*Noise spectrum of the intensity difference.* S_0 *is the shot noise for a beam of intensity* $I_1 + I_2$, $\omega_c = \frac{2\gamma'}{\tau}$ *is the cavity bandwith. Curve (a) corresponds to no internal losses* ($\mu = 0$), *curve (b) to equal transmission and losses* ($\mu = \gamma$), *while the total losses* γ' *are the same in a) and b).*

The fluctuation spectrum of the phase difference between the twin fields can be calculated in the same way. Setting :

$$s = \frac{1}{\sqrt{2}} (q_1 - q_2) \tag{39}$$

one finally gets :

$$S_s (\omega) = 1 + \frac{4\gamma\gamma'}{\omega^2 \tau^2} \tag{40}$$

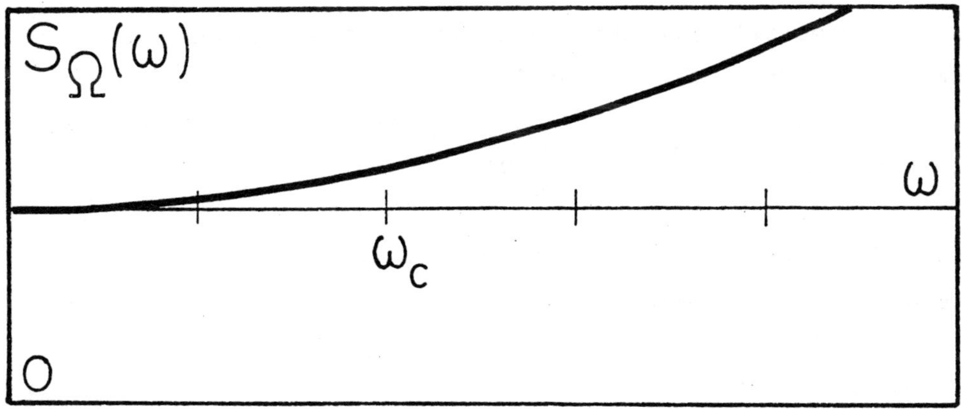

Figure 5 : Noise spectrum of the beat frequency between the two signal fields. The zero-frequency limit correspond to the Schawlow-Townes limit associated with phase diffusion.

This expression diverges when ω tends to 0, which is characteristic of a phase diffusion process. One can deduce the fluctuation spectrum $S_\Omega (\omega)$ of the beat frequency between the two signal modes :

$$\Omega = \omega_1 - \omega_2 = \frac{1}{\sqrt{2}\, \bar{\alpha}_1} \dot{s}$$

(where $\bar{\alpha}_1 = \bar{\alpha}_2$ is the mean signal field). One thus gets :

$$S_{\Omega}(\omega) = \frac{1}{2\bar{\alpha}_1^{2\,out}} \omega^2 S_s(\omega)$$

$$= \frac{1}{2I_1^{out}} \left(\frac{4\gamma\gamma'}{\tau^2} + \omega^2 \right) \qquad (41)$$

The beat frequency noise spectrum is displayed in Fig.(5) : the zero frequency limit corresponds to the Schawlow-Townesfluctuations associated with phase diffusion (Sargent et al. 1974), while the frequency dependent part describes the vacuum fluctuations back reflected onto the output mirror (Yamamoto et al. 1986).

We now give the results for the other quantities : fluctuation of the sum of the phases :

$$q = \frac{1}{\sqrt{2}} (q_1 + q_2) \qquad (42)$$

and of the sum of the intensities :

$$p = \frac{1}{\sqrt{2}} (p_1 + p_2) \qquad (43)$$

In contrast to r and s, these quantities involve the pump fluctuations. In our experimental case, the cavity finesse for the pump is much lower than for the signals ($\gamma_0' \gg \gamma_1', \gamma_2'$) and we will only consider the noise at frequencies small compared to γ_0'/τ . Consequently, we shall neglect $i\omega\tau$ in the Fourier transform of Eq.(27-c) yielding :

$$\gamma_0' \, \delta\alpha_0 = -\sqrt{\gamma_0' \, \gamma'(\sigma-1)} \, (\delta\alpha_1 + \delta\alpha_2) + \sqrt{2\gamma_0} \, \delta\alpha_0^{in} + \sqrt{2\mu_0} \, \beta_0^{in} \tag{44}$$

(this is analogous to the usual adiabatic elimination procedure).

Using the same method as before one finally gets :

$$S_q(\omega) = \frac{\omega^2 \tau^2 + 4\gamma'^2(\sigma^2 - 1) + 4\gamma'\mu}{\omega^2 \tau^2 + 4\gamma'^2 \sigma^2} \tag{45}$$

$$S_p(\omega) = \frac{\omega^2 \tau^2 + 4\gamma'^2 (\sigma-1)^2 + 4\gamma\gamma'}{\omega^2 \tau^2 + 4\gamma'^2 (\sigma-1)^2} \tag{46}$$

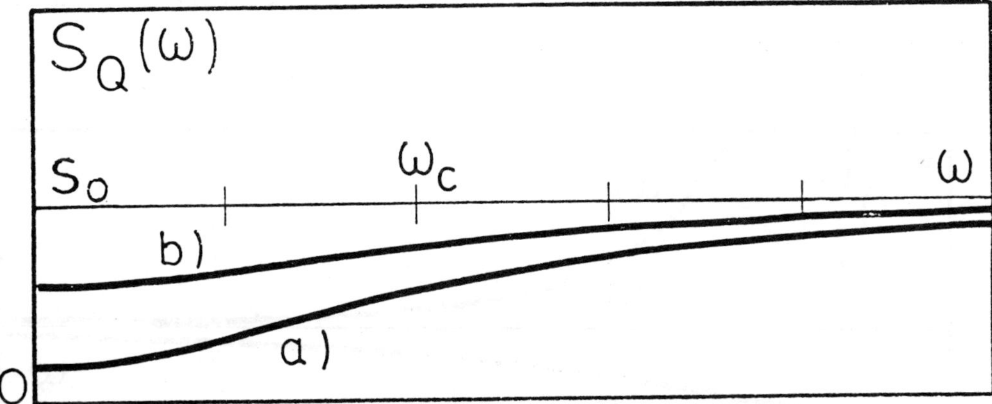

Figure 6 : Noise spectrum on the sum of the output field phases. The ratio σ between the pump field amplitude and the threshold pump field amplitude is chosen equal to 1.1 curve ; a) coresponds to no internal losses (μ = 0), curve b) to equal transmission and losses ($\mu = \gamma$ with γ' kept constant).

These quantities now depend on σ, that is, on the pump power. For the sum of the phases, complete noise suppression is obtained for $\gamma'=\gamma$ and at threshold (σ=1) , and noise reduction is expected inside the cavity bandwidth. This result is the equivalent of the squeezing observed in the degenerate OPO on the in-quadrature component. When γ' is larger than γ, the noise reduction factor is decreased by the same amount as in the intensity difference case (Fig.5).

The noise spectrum $S_{I_1+I_2}$ for the sum of intensities is equal to :

$$S_{I_1+I_2}(\omega) = S_0 \; S_p \; (\omega) \tag{47}$$

(S_0 is the same shot noise as in Eq.(37)). At zero frequency, $S_{I_1+I_2}$ diverges close to threshold ($\sigma \to 1$) and goes down to the shot noise limit when $\sigma \to \infty$. As well as the in-phase component in the degenerate OPO, this component exhibits amplified fluctuations.

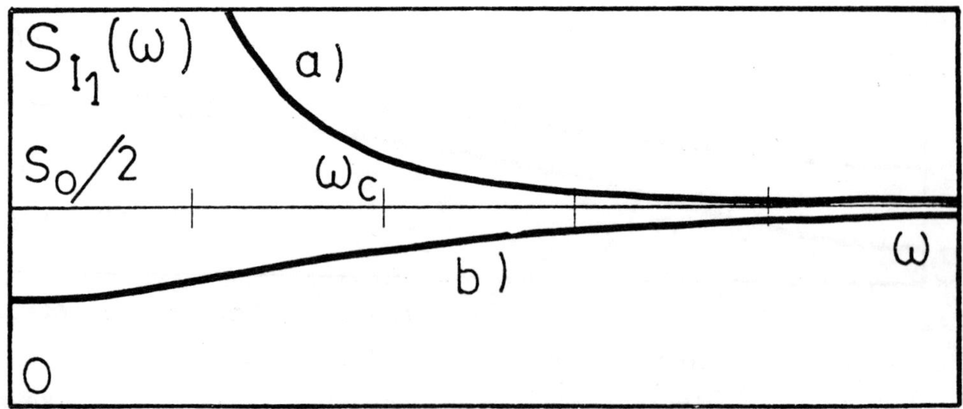

Figure 7 : Noise spectrum of the intensity of one output field, in the case of zero internal losses (μ=0). Curve a) corresponds to σ=1.1, curve b) to σ=5.

However, an interesting feature appears on the fluctuations of a single beam intensity.

$$S_{I_1}(\omega) = \frac{S_0}{2}\left(1 - \frac{8\gamma\gamma'^3\ \sigma(\sigma-2)}{(\omega^2\ \tau^2\ +4\gamma'^2\ )(\omega^2\ \tau^2 + 4\gamma'^2(\sigma-1)^2)}\right) \qquad (48)$$

where $\frac{S_0}{2}$ is the shot noise on one of the twin beams. One can see (in Fig. 7) that starting from above shot noise near threshold, $S_{I_1}(\omega)$ goes below the shot noise for $\sigma > 2$ in a bandwidth of the order of the cavity bandwidth. At best, for $\omega=0$, and $\sigma\to\infty$:

$$S_{I_1}(0) = \frac{S_0}{2}\left(1 - \frac{\gamma}{2\gamma'}\right) \qquad (49)$$

If there are no internal losses, the shot noise is reduced by a factor 2.

4.5 Non-balanced case

In the general case of unequal transmission and loss coefficients, it is possible to analytically solve the equations giving the Fourier components of the amplitude and phase noise $\tilde{p}_1(\omega)$, $\tilde{q}_1(\omega)$.

We will not give here the lengthy expressions for $S_{p_1}(\omega)$ and $S_{q_1}(\omega)$ but rather focus on their physical content. We will restrict ourselves to the discussion of the noise spectrum on the intensity difference $S_I(\omega)$ in the case where the pump cavity finesse is much smaller than the signal cavity finesse ($\gamma_0 \gg \gamma_1, \gamma_2$) which corresponds to our experimental situation.

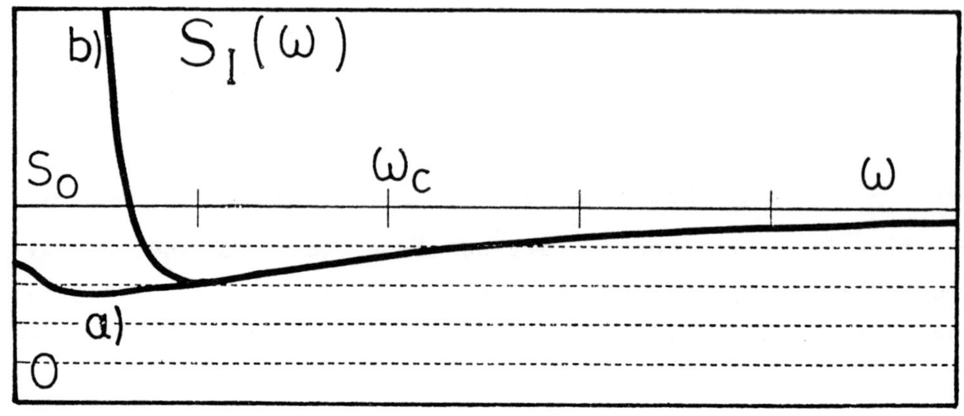

Figure 8 : Noise spectrum of the intensity difference in the non-balanced case. For the displayed curves, the losses coefficients and transmission are slightly different (γ_1 = 0,55% , γ_2 = 0,45%, μ_1 = 0,48%, μ_2 = 0,52%) *so that the twin beam intensity mean values differ by 15%. The pump amplitude parameter σ is equal to 1.1. Curve a) corresponds to a shot noise limited pump beam. Curve b) corresponds to the pump beam with 30dB extra noise at zero frequency decaying exponentially to zero at high frequency with a half width of 0.15* ω_c .

The main consequence of the unbalance is that the pump intensity fluctuations are now coupled into $S_I(\omega)$, resulting in an increased noise at any frequency on the whole spectrum.

We consider the case in which μ_1 and γ_1 are not very different from μ_2 and γ_2. Then we find that the added noise is only important very close to threshold ($\sigma \approx 1$) and in the very low frequency part of the spectrum. Fig.(8-a) gives $S_I(\omega)$ for $\sigma = 1.1$, and for loss coefficients giving 15% difference between the mean intensities of the two beams, and assuming that the pump field is shot noise limited at any frequency :

one notices a moderate increase of the noise at low frequencies. Fig. (8-b) takes into account extra pump noise at low frequency, which corresponds to a more realistic situation : one then observes a sharp rise in the noise for $\omega \approx 0$. Of course, if the low frequency pump intensity fluctuations are too large, our simple linearized approach for the fluctuations dynamics breaks down, and multiplicative noise may appear. But the general conclusion that $S_I(\omega)$ is increased because of unbalance only in a limited range of low frequencies remains certainly valid.

5. EXPERIMENT

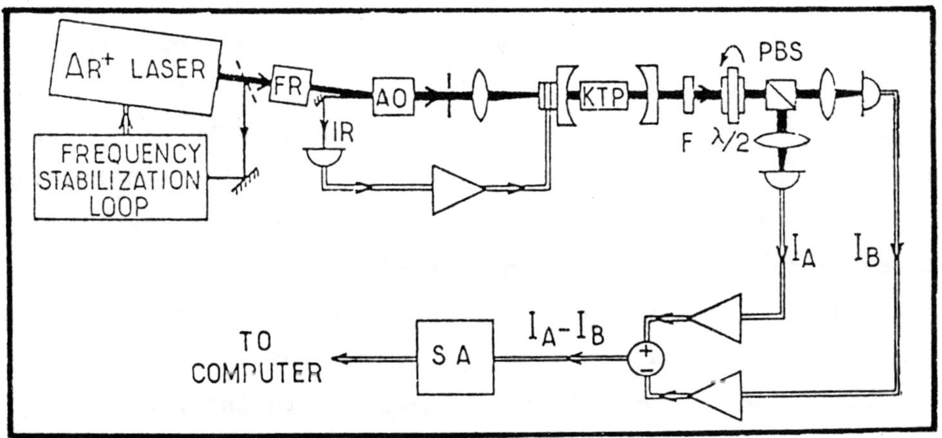

Figure 9 : Experimental set-up. FR : Faraday rotator ; AO : acousto-optic modulator ; F : filter ; PBS : polarizing beam-splitter ; SA : spectrum analyzer.

We use a two-mode parametric oscillator operated above

threshold to generate high intensity twin beams (Heidmann et al. 1987). The aim of the experiment is to study the fluctuation spectrum on the intensity difference between the twin beams. We will show that significant reduction below the shot noise limit is observed over a broad range of frequencies.

Figure (9) shows the experimental set-up. The OPO is pumped by a single mode Ar^+ ion laser at 528nm stabilized on an external Fabry-Perot cavity (residual frequency jitter of 50kHz). A Faraday rotator (FR) and an acousto-optic modulator (AO) are used to optically isolate the laser from the strong back reflected light coming from the OPO. The pump light is then focused into the parametric medium, which is a 7mm long, type II phase-matched KTP ($KTiOPO_4$) crystal, inserted in an optical cavity of length 17mm. The input mirror, with a 2cm radius of curvature, is highly reflecting for the infrared signal and idler beams and transmitting for the pump beam. The output mirror is flat and transmits 0.8% of the infrared light and a large part of the green light. Consequently, the cavity finesse is high for the signal and idler fields and low for the pump field. The four OPO oscillation conditions (energy conservation, phase matching condition, cavity resonance conditions for both signal and idler) are only fulfilled for a discrete series of cavity length values. In practice, oscillation occurs only in very small length intervals (a few nanometers) around these values. This is the cause of the well-known high sensitivity of the OPO to vibrations (Smith 1973). Thus the OPO length has to be actively stabilized by electronic feedback so that it delivers a nearly constant output intensity. For this purpose, the OPO output is monitored on the weak counterpropagating infrared beam (IR) which is transmitted back through the input mirror and not deflected by the acousto-optic modulator, and compared to a stable reference level. As a consequence, the OPO length is stabilized on the side of one of such small oscillation domains, corresponding to a slight mismatch between the OPO frequencies and the cavity eigen frequencies.In some experimental conditions, we can observe a bistable behaviour as a function of cavity length, but one can always choose for stabilization the right side of the oscillation interval, which does not yield a bistable regime. Above threshold (80mW of green light), the OPO emits two cross-polarized twin beams (with intensities of a few milliwatts for 200mW of green light). The emission wavelengths, λ_1 = 1.048 µm and λ_2 = 1.067 µm, are determined by the collinear phase matching conditions. The remaining transmitted pump beam is stopped by a filter. The twin beams are separated by a polarizing beamsplitter (PBS) and then focused on two InGaAs photodiodes which have quantum efficiencies of 90%. All surfaces encountered by the two infrared beams are antireflection coated. The two photocurrents are amplified, and then subtracted using a 180° power combiner. The noise on the resulting difference is monitored by a spectrum analyzer connected to a computer for data analysis.

The characteristics of the detection channels have been carefully checked : the imperfections of the polarizing beamsplitter are less than 1 % ; the amplifier voltage gains are matched within 1 %. The overall common mode rejection between the two channels has been measured by modulating the pump beam at a frequency of 10 MHz and measuring the corresponding coherent peak reduction on the spectrum analyzer. The result of this measurement is 25 dB.

A key point for the reliability of such an experiment is the calibration of the shot noise level. As a first test, we have used a rotating half-wave plate inserted in front of the polarizing beamsplitter (labeled $\lambda/2$ in Fig.9). The two fields E_1 and E_2 emitted by the OPO undergo a polarization rotation of 2θ in the half-wave plate, where θ is the angle between the axes of the plate and of the polarizer. The two fields E_A and E_B respectively transmitted and reflected by the polarizing beam splitter are :

$$E_A = \cos 2\theta \; E_1 - \sin 2\theta \; E_2 \tag{50-a}$$

$$E_B = \sin 2\theta \; E_1 + \cos 2\theta \; E_2 \tag{50-b}$$

When $\theta = 0^\circ$ (modulo 45°), the half-wave plate plays no role and the measured signal is the difference between the twin beam intensities. When $\theta = 22.5^\circ$ (modulo 45°), the system half-wave plate and polarizing beamsplitter acts like an usual 50% beamsplitter. Since in our experimental conditions, the beat frequency between the twin fields is about 5 THz, the crossed terms between the two modes do not appear in the observed frequency range. Consequently, the measured signal gives the shot noise level for a beam of intensity $I_1 + I_2$. One can show from Eqs.(50) that the noise power spectrum $S_\theta(\Omega)$ for the signal $I_A - I_B$ varies sinusoidally as a function of the angle θ :

$$S_\theta(\Omega) = S_I \; (\Omega) \; \cos^2 4\theta + S_0 \; \sin^2 4\theta \tag{51}$$

Fig.10 shows the variation of $S_\theta(\Omega)$ recorded at a fixed frequency $\Omega/2\pi$ = 8MHz. One observes a strong modulation of the noise level with the expected periodicity of 45° : the noise level at 0° is about 30% lower than the one at 22.5°.

As a second test, we have used a cw YAG laser to yield an independent characterization of the shot noise level. We have checked that the YAG laser was shot noise limited at frequencies higher than 2MHz. We have then measured the noise levels of the OPO and YAG with equal mean intensities. From this test, we can assert that the upper level of Fig. 22 (θ=22.5°) coincides within 1% with the shot noise level (dashed line in Fig.10).

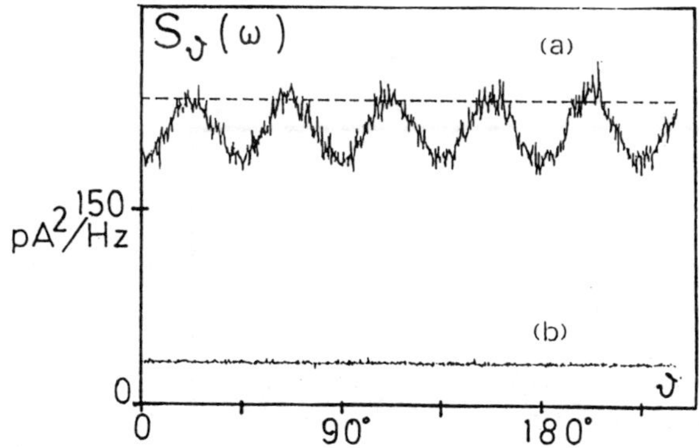

Figure 10 : a) Variation of the measured noise power on the intensity difference S_θ (Ω) *as a function of* θ *for* $\Omega/2\pi$=8MHz *(expressed in terms of photodiode current noise) ; b) input noise level equivalent to the whole electronic noise. The dashed line gives the shot noise level at the same frequency of a YAG laser having the same intensity. Scan time for a) and b) is 50s, without videofilter.*

Fig.11 gives the noise reduction factor R(Ω) which is the ratio of the "squeezed" noise spectrum, recorded at θ = 0°, to the shot noise spectrum, θ = 22.5° (both spectra have been corrected from electronics noise). The curve is clearly below 1 over a broad frequency range. A maximum noise reduction of 30% ± 5% is observed at a frequency of 8MHz. The noise reduction is better than 15% from 3 to 13MHz.

In the low frequency domain, the noise increases because the large extra noise on each beam is not completely rejected in the difference process. Moreover, the mean intensities are not exactly equal, which we attribute to a slight difference in the losses for the two infrared beams. This is a cause for additional fluctuations to be coupled back into the measured signal, as discussed in the theoretical § 4.

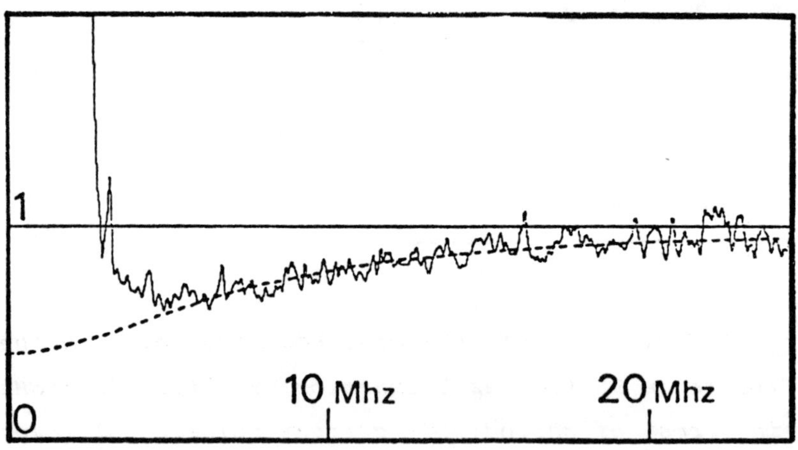

Figure 11 : Experimental noise reduction factor $R(\Omega)$ spectrum. It is obtained by recording 3 spectra : the "squeezed" noise ($\theta = 0$) $S(\Omega)$, the shot noise ($\theta = 22,5^\circ C$) $N(\Omega)$ and the electronic noise $E(\omega)$, $R(\Omega) = (S(\Omega) - E(\omega))/(N(\omega) - E(\omega))$. The dashed line is a Lorentzian fit of the experimental points.

At high frequencies, $R(\Omega)$ is seen to go to 1 : the

noise of $I_1 - I_2$ rises back to the shot noise for frequencies higher than the cavity bandwidth. The dashed line shows a Lorentzian fit of the experimental spectrum (for frequencies higher than 5 MHz). The extrapolated value at $\omega=0$ is about 50%, which is in agreement with the value one would expect from the effect of the internal losses.

6. APPLICATIONS OF TWIN BEAMS

The OPO provides high intensity beams made of correlated photons. This feature appears very promising to reduce quantum noise in several experimental configurations, and therefore to enhance the sensitivity of quantum noise limited measurements. In this paragraph, we will outline two such applications : generation of light beams below the shot noise limit and measurement of very small absorptions.

6.1 Reduction of intensity noise below shot noise on a single light beam

As suggested by several authors (Saleh et al. 1985, Jakeman et al. 1986, Haus et al. 1986, Yuen 1986, Yamamoto et al. 1987), correlated beams may be used for "intensity squeezing" on a single beam, via electronic intensity correction. Two possible experimental schemes are shown in Fig.(12). In both of them, the intensity of beam B_2 is monitored by a photodiode.In the first scheme (Fig.(12-a)), such an information is used to modify the transmission of an intensity modulator, in order to correct the pump intensity. In the second one (Fig. (12-b)), the same information is used directly to correct the twin beam I_1 by the same technique. The main difference between the two schemes is that there is a loop process in the first one which acts in a selfconsistent way, which is not the case in the second one.

Let us now consider the first technique in a more quantitative way. From Eqs.(27) and(28), we can write the in-phase fluctuations (in Fourier space) in the balanced case as :

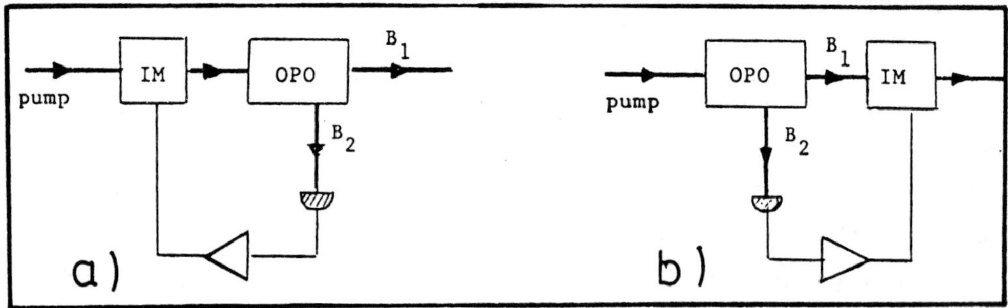

Figure 12 : Two possible schemes of intensity correction : the intensity of beam B_2 is used in a) to correct the pump intensity by acting on an intensity modulator (IM) (feedback) and in b) to correct the intensity of beam B_1 feedforward).

$$\tilde{p}_1^{out} = \pi_1 + a\,\tilde{p}_0^{in} \tag{52-a}$$

$$\tilde{p}_2^{out} = \pi_2 + a\,\tilde{p}_0^{in} \tag{52-b}$$

where $a\,\tilde{p}_0^{in}$ is the part of fluctuations coming from the pump fluctuations and π_1 and π_2 the remaining part due to other noise sources. The electronic feedback loop produces a signal proportional to p_1^{out} which is used to react on the pump beam intensity. Let t be the amplitude transmission factor associated with the amplitude modulator. The pump field amplitude α_0^t transmitted by the modulator is then :

$$\alpha_0^t = t\alpha_0 + \sqrt{1-t^2}\ \alpha_0^n \tag{53}$$

where α_0^n is the noise fluctuating field entering the system because of the losses introduced by the intensity modulator. Due to the feedback loop, the transmission t depends on the intensity fluctuations of beam 1, p_1^{out}, according to :

$$t = \bar{t} - \frac{G}{\bar{\alpha}_0} p_1^{out}$$

where $\bar{t}$ is the amplitude transmission factor without feedback and G the effective loop gain factor, assumed here, for the sake of simplicity, to be real and frequency independent.

Then :

$$\alpha_0^t = \bar{t}\,\bar{\alpha}_0 + \bar{t}\,\delta\alpha_0 - Gp_1^{out} + \sqrt{1-\bar{t}^2}\ \alpha_0^n \tag{54}$$

The real part of such fluctuations must then be inserted in eq. (52) as the value of the pump intensity fluctuations impinging on the OPO, which gives the following self-consistent value for $\tilde{p}_1^{out}$:

$$\tilde{p}_1^{out} = \left[\pi_1 + a(\bar{t}\,\tilde{p}_0^{in} + \sqrt{1-\bar{t}^2}\ \tilde{p}_0^n)\right] \frac{1}{1+aG} \tag{55}$$

This value is now inserted in eq. (54) to give the effective pump fluctuations, which yields the following expression for the in-phase fluctuations on beam B_2 :

$$\tilde{p}_2^{out} = \pi_2 - \frac{Ga}{1+Ga}\pi_1 + \frac{a}{1+Ga}\left(\bar{t}\ \tilde{p}_0^{in} + \sqrt{1-\bar{t}^2}\ \tilde{p}_0^{n}\right) \quad (56)$$

When the loop gain goes to infinity, this expression simply reduces to :

$$\tilde{p}_2^{out} = \pi_2 - \pi_1 \quad (57)$$

As expected, in the high gain limit, the feedback loop transfers on beam B_2 the open loop fluctuations on $p_1^{out} - p_2^{out}$, i.e. on the intensity difference. As a result, the noise spectrum on the feedback corrected intensity is :

$$\left[S_{I_2}(\omega)\right]_{feedback} = \left[S_{I_1-I_2}(\omega)\right]_{openloop} = S_0\ \frac{\omega^2\ \tau^2\ +\ 4\mu\gamma'}{\omega^2\ \tau^2\ +\ 4\gamma'^2} \quad (58)$$

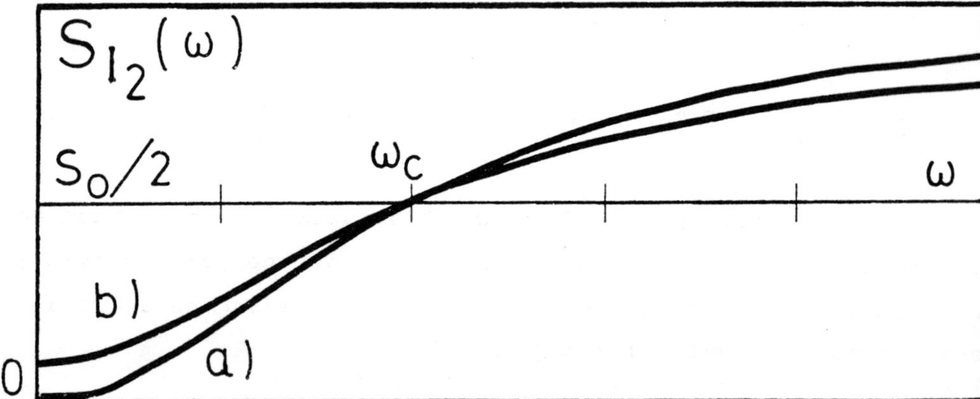

Figure 13 : Intensity noise spectrum of beam B_2 after electronic correction. So/2 is the shot noise of a classical beam having the same mean intensity. The mean transmission of the intensity modulator is 80% and there are no internal losses in the OPO. a) feedback technique, b) feedforward technique.

Let us stress that the shot noise level S_0, defined as previously (Eq.(7)), corresponds to a beam of intensity $I_1 + I_2 = 2I_1$. In formula (58) displayed in Fig. (13-a), the background reached for $\omega \to \infty$ is therefore twice as large as the standard shot noise for a beam of intensity I_1. It is thus possible to reduce the intensity noise on one beam, but <u>only when the noise reduction factor on $I_1 - I_2$ is larger than 50%</u>.

This is due to the fact that, when one monitors the fluctuations of beam B_1 to correct for the fluctuations of beam B_2, the feedback loop actually adds noise if the quantum fluctuations are uncorrelated.

Let us now consider the problem of the "feed forward" correcting device. In this case, the intensity fluctuations on beam B_1 are used to act on a variable transmitter inserted on beam B_2. The transmitted field amplitude α_2^t through this device has an expression similar to Eq. (53):

$$\alpha_2^t = t\alpha_2 + \sqrt{1-t^2}\ \alpha_2^n \qquad (59)$$

Here the extra noise term α_2^n is the unescapable counterpart of a variable transmission acting on beam B_2. Then :

$$\alpha_2^t = \bar{t}\ \bar{\alpha}_2 + \bar{t}\ \alpha_2 - Gp_1^{out} + \sqrt{1-\bar{t}^2}\ \alpha_2^n \qquad (60)$$

Taking again G real and frequency independent, one finds that the in-phase fluctuations of beam B_2 after the variable transmitter are given by :

$$\tilde{p}_2^t = \bar{t}\pi_2 - G\pi_1 + B\tilde{p}_0^{in}\ (\bar{t} - G) + \sqrt{1-\bar{t}^2}\ \alpha_2^n \qquad (61)$$

In order to transfer on beam B_2 the noise existing on $I_1 - I_2$, like in the previous case, we need to choose a precise value of the effective loop gain $G = \bar{t}$, in which case the noise spectrum of the feedfordward corrected intensity I_2 is given by :

$$\left[S_{I_2}(\omega)\right]_{feedforward} = S_0\ \bar{t}^2\ \frac{\omega^2\ \tau^2\ +\ 4\gamma'\mu}{\omega^2\ \tau^2\ +\ 4\gamma'^2} + \frac{S_0}{2}\ (1-\bar{t}^2) \qquad (62)$$

This final formula, displayed in Fig. (13-b), can be compared to the formula obtained in the feedback configuration : in the two cases, we see that an open loop noise reduction of 50% at least is required to reduce the intensity noise on I_1, for the reason given previously. But one may notice the following differences between the two situations :

(i) The gain value necessary to reduce the noise is precisely fixed in the second case, and needs only to be high in the first one. This makes the second set-up very sensitive to the electronic gain fluctuations.

(ii) the noise entering the system via the variable attenuator does not matter in the feedback case, whereas it lowers the squeezing factor in the second case. A transmission factor close to unity is therefore required in the feedforward case.

(iii) If one takes into account the time delays in the correcting system (neglected in this first simple approach), the feedback system may very easily lead to spurious oscillation, which is never the case in the feedforward system.

6.2 Measurement of weak absorption beyond the shot noise limit

Intensity correlated beams are ideally suited to reduce the noise floor in absorption measurements. The principle of such an experiment is extremely simple (Fabre et al. 1986) : an absorbing medium having an absorption coefficient $\epsilon(\omega)$ resonant at a given frequency ω_a is inserted on the beam B_1. The phase-matching condition of the OPO is varied (by temperature or angle tuning) so that the frequency ω_1 varies around ω_a, and the absorption dip is measured on the intensity difference $I_1 - I_2$ as a function of frequency. As the background noise power on such a signal is reduced by a factor $\frac{\mu}{\gamma'}$, with respect to shot noise, this technique allows us to measure a minimum absorption coefficient $\epsilon(\omega_a)$ smaller by a factor $\sqrt{\frac{\mu}{\gamma'}}$. This method will be useful for example in measurements where increasing the laser intensity is impossible (biological samples for instance).

But the simple technique described above, as others of the same kind, will always be hampered by the unavoidable low frequency excess noise, which will presumably prevent any high sensitivity measurement at zero frequency. Modulation techniques are then needed, which transfer the signal to be measured in a frequency range where the noise is minimum (Gehrtz et al. 1985).

However, phase or amplitude modulation at a frequency ω adds quantum noise (Yurke at al. 1987a) : this can be simply understood by recalling that any modulation techniques couples modes having frequency differences equal to ω. Vacuum fluctuations of empty modes distant by ω from filled modes are then coupled back into the system, giving rise to an increase of fluctuations. One then needs more elaborate techniques such as multiple frequency squeezing in order to reduce the noise in phase-modulation experiments with the help of squeezed states (Gea-Banacloche 1987, Yurke et al. 1987b). In intensity modulation experiments as well, the modulator introduces periodic losses having a non zero mean value, and therefore adds a significant amount of quantum noise.

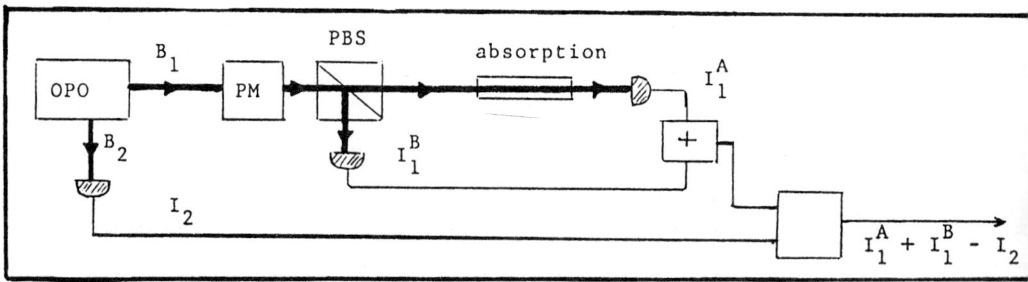

Figure 14 : Modulation scheme allowing to transfer the absorption signal at given frequency ω without adding quantum noise, by using a polarization modulator (PM) and a polarizing beam-splitter (PBS).

It is nevertheless possible to modulate the signal without adding noise : this is done for example by modulating not the beam itself but the absorption coefficient $\epsilon(\omega)$, by different means such as an oscillating Stark shift or level population pulsation, etc.. We present in Fig. (14) another possibility : let us insert on beam B_1 a polarization modulator and polarizing beam-splitter PBS. In such a way, we get the following values for the two mean intensities $\bar{I}_1^A$ and $\bar{I}_1^B$ going out of the device :

$$\begin{cases} \bar{I}_1^A = \cos^2\omega t \; \bar{I}_1 \\ \bar{I}_1^B = \sin^2\omega t \; \bar{I}_1 \end{cases} \tag{63}$$

Of course, such a system adds quantum noise, because it couples the output to the vacuum fluctuations on the input mode β_1 having a polarization orthogonal the beam B_1 and the same frequency. More precisely the output fields α_1^A and α_2^A are given by :

$$\begin{cases} \alpha_1^A = \cos\omega t \; \alpha_1 + \sin\omega t \; \beta_1 \\ \alpha_1^B = -\sin\omega t \; \alpha_1 + \cos\omega t \; \beta_1 \end{cases} \tag{64}$$

then :

$$\begin{cases} I_1^A = |\alpha_1^A|^2 = \cos^2\omega t \; I_1 + \sin 2\omega t \sqrt{I_1} \; \beta_1 \\ I_1^B = |\alpha_1^B|^2 = \sin^2\omega t \; I_1 - \sin 2\omega t \sqrt{I_1} \; \beta_1 \end{cases} \tag{65}$$

An absorbing medium is then inserted on beam B_1^A and one monitors the absorption dip on the signal $I_1^A + I_1^B - I_2$. When there is no absorption the extra noise added by the modulation cancels when one measures the sum of intensities $I_1^A + I_1^B$. Therefore the noise on $I_1^A + I_1^B - I_2$ is the same as the noise on $I_1 - I_2$ without modulation : since no photon is lost in the modulation process, no noise is added. At frequency ω_a, one measures the signal :

$$S = \bar{I}_1 \cos^2\omega t \; \epsilon(\omega_a) \tag{66}$$

having a Fourier component $(\bar{I}_1/2)\;\epsilon(\omega_a)$ at frequency 2ω. If we choose 2ω equal to the frequency of maximum noise reduction, we can take full advantage of such a reduction to obtain absorption measurement beyond shot noise.

ACKNOWLEDGEMENT

This work has been done with the support of EEC Stimulation Action Grant Number ST2J0278C.

REFERENCES

Björk, G. and Yamamoto, Y. 1988 Phys. Rev. A 37 125.

Burnham, D.C. and Weinberg, D.L. 1970 Phys. Rev. Lett. 25 84.

Collett, M.J. and Gardiner, C.W. 1984 Phys. Rev. A 30 1386.

Fabre, C., Giacobino, E., Reynaud, S., Debuisschert, 1986 T. SPIE 70 ECOOSA'86 (Florence) 489.

Friberg, S., Hong, C.K. and Mandel, L. 1985 Phys. Rev. Lett. 54 2011.

Gardiner, C.W. 1983 "Handbook of Stochastic Methods" (Springer Verlag).

Gea-Banacloche, J. and Leuchs, G. 1987 J. Opt. Soc. Am. B 4 1667.

Gehrtz, M., Bjorklund, G.C. and Whittaker, E.A. 1985 J. Opt. Soc. Am. B 2 1510.

Graham, R. 1973 Springer Tracts in Modern Physics (Springer Verlag, Berlin) vol. 66.

Graham, R. 1984 Phys. Rev. Lett. 52 117.

Haus, H. and Yamamoto, Y. 1986 Phys. Rev. A 34 270.

Heidmann, A., Horowicz, R.J., Reynaud, S., Giacobino, E., Fabre, C. and Camy, G. 1987 Phys. Rev. Lett. 59 2555.

Hong, C. and Mandel, L. 1985 Phys. Rev. A 31 2409.

Jakeman, E. and Rarity, J.G. 1986 Optics Comm. 59 219.

Lane, A.S., Reid, M.D. and Walls, D.F. 1988 Phys. Rev. A to be published.

Louisell, W.H., Yariv, A. and Siegmann, A.E. 1961 Phys. Rev. 124 1646.

Lugiato, L.A. and Strini, G. 1982 Optics Comm. 41 67.

Milburn, G. and Walls, D.F. 1981 Optics Comm. 39 401.

Mollow, B.R. and Glauber, R.J. 1967 Phys. Rev. 160 1076 and 1097.

Mollow B.R. 1973 Phys. Rev. A 8 2684.

Reynaud, S. 1987 Europhysics Lett. 4 427.

Reynaud, S., Fabre, C. and Giacobino, E. 1987 J. Opt. Soc. Am. B 4 1520.

Saleh, B. and Teich, M. 1985 Optics Comm. 52 429.

Sargent, M., Scully, M., Lamb, W.1974 Laser Physics Addison-Wesley, Reading (USA).

Savage, C.M. and Walls, D.F. 1987 J. Opt. Soc. Am. B 4 1514.

Slusher, R.E., Grangier, P., La Porta, A., Yurke, B. and Potasek, M. 1987 Phys. Rev. Lett. 59 2566.

Smith, R.G. 1973 "Optical Parametric Oscillators", in Laser Handbook, Arecchi, F.T. and Schultz-Dubois E. Ed. North Holland Amsterdam.

Stoler, D. 1974 Phys. Rev. Lett. 33 1397.

Takahasi, H. 1965 Adv. Comm. System 1 227.

Wu, L.A., Kimble, H.J., Hall, J.L. and Wu, L. 1986 Phys. Rev. Lett. 57 2520.

Yamamoto, Y., Machida, S. and Nilsson, O. 1986 Phys. Rev. A 34 4025.

Yamamoto, Y., Machida, S. Imoto, N., Kitagawa, M. and Björk, G.1987 Soc. Am. B 4 1645.

Yuen, H.P. 1986 Phys. Rev. Lett. 56 2176.

Yurke, B. 1984 Phys. Rev. A 29 408.

Yurke, B. and Whittaker, E.A. 1987a Opt. Lett. 12 236.

Yurke, B., Grangier, P. and Slusher, R.E. 1987b J. Opt. Soc. Am. B 4 1677.

NONCLASSICAL EFFECTS IN PARAMETRIC DOWNCONVERSION

J G RARITY AND P R TAPSTER

1 INTRODUCTION

The mixing of three electromagnetic waves within a medium with a second order refractive non-linearity can lead to several different parametric processes. The semiclassical theory of these mixing effects is well reviewed by Yariv (1976). The semiclassical methods hide the fundamental quantum nature of the interaction. In parametric downconversion the quantum effects arise because two photons are created from one pump photon. These photons satisfy the necessary conditions of momentum (phase) and energy conservation within the non-linear medium. In non-degenerate downconversion separable photons (either by angle or by polarisation) are created. Burnham and Weinberg (1970) demonstrated that the two photons are created nearly simultaneously, using photon counting coincidence techniques, while Mollow (1973) developed a quantum mechanical treatment allowing the two photon coincidence detection probability to be calculated. Mollow suggested that the time separation of the two photons would be related to the coherence time of the illuminating radiation. This has since been disproved by more recent coincidence experiments with time resolutions of a hundred picoseconds (Friberg et al 1985).

Using non-degenerate type I downconversion two identical trains of photons can be selected by apertures satisfying the phase matching conditions. Detection of photons in one train can be used to modify the quantum statistics of the other train to produce antibunched and/or sub-poissonian light (section 2). We report here on two experiments of this type. The first employs single photon detection techniques and a fast optical shutter to create antibunched and slightly sub-poissonian light. The second uses analogue detection and feedback to produce sub-poissonian light at picowatt power levels with a Fano factor of 0.78 over a limited bandwidth. This, to our knowledge, is the lowest post-detection Fano factor yet reported. Sub-poissonian light is potentially useful in finite dose transmission measurements and the high degree of temporal coincidence of the downconverted photon pairs can be exploited in absolute measurement of detector quantum efficiency (section 3).

Up until recently, measurement of the degree of temporal coincidence has been apparatus limited. Recent theory (Fearn and Loudon, 1987;Ou et al, 1987; Prasad et al, 1987) and experiment (Hong and Mandel, 1987; 1988) has shown that recombining indistinguishable photon pairs in a beamsplitter leads to a fourth order interference effect which allows a direct measure of the photon overlap. Such experiments allow measurement of sub-picosecond time delays at the single photon level. We show here (section 4) that the detailed shape of the effect can be related to the fourier transform of the photon bandwidth determined by the phase matching uncertainty and aperture size. Experimental results are presented which confirm this theory. The minimum photon length is limited by the bandwidth of the crystal non-linearity coupled with the detailed physics of the parametric downconversion process (Pike and Sarkar, 1988).

2 ANTIBUNCHING AND SUB-POISSONIAN LIGHT SOURCES

2.1 Phase matching conditions for non-degenerate parametric downconversion

For non-degenerate parametric downconversion (Yariv, 1976) a non-linear uniaxial crystal is used. The crystal is aligned with its optic axis at an angle ϕ with an incident short wavelength laser beam. About 10^{-8} of the beam is converted into pairs of longer wavelength photons which appear in a cone surrounding the laser beam. The photon pairs satisfy the conditions of energy and wave vector conservation in the crystal. Given an incident photon of wave vector $\vec{k}_0$ and angular frequency ω_0, and downconverted photons with wave vectors $\vec{k}_1$, $\vec{k}_2$ and angular frequencies ω_1, ω_2, these conditions can be written as

$$\vec{k}_0 = \vec{k}_1 + \vec{k}_2 \tag{1}$$

$$\omega_0 = \omega_1 + \omega_2 \tag{2}$$

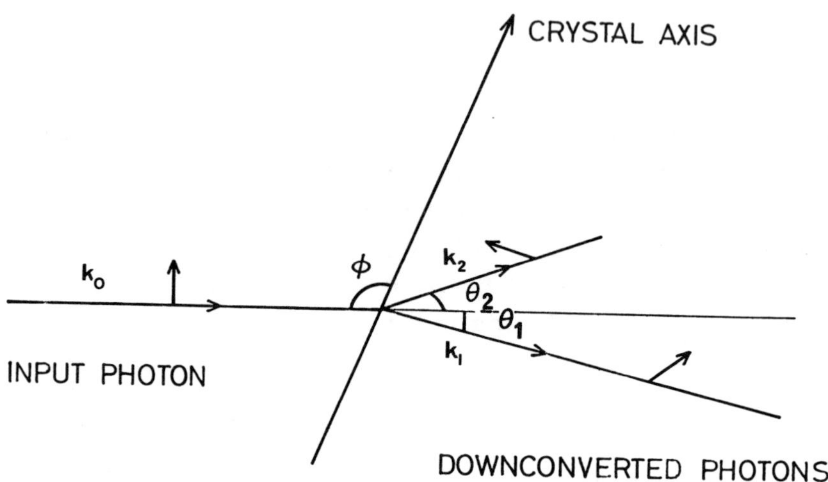

Figure 1: The geometry of type one phase matching.

The geometry illustrated in figure 1 depicts type I phase matching, where the incoming beam is polarised in the plane of the crystal optic axis, and the downconverted light is polarised perpendicular to this. In this case equation 1 can be rewritten

$$\omega_0 n_0 = \omega_1 n_1 \cos\theta_1 + \omega_2 n_2 \cos\theta_2$$
$$\omega_1 n_1 \sin\theta_1 = \omega_2 n_2 \sin\theta_2 \tag{3}$$

where n_1 and n_2 are the ordinary refractive indices of the crystal at frequencies ω_1 and ω_2, and n_0 is given by

$$n_0 = n_e \sin\phi + n_o \cos\phi \tag{4}$$

where n_o is the ordinary and n_e is the extraordinary refractive index of the crystal at ω_0. As a consequence of equations 2-4 the crystal must be negatively uniaxial ($n_e < n_o$) and the cone angle is maximised by choosing $\phi = 90°$. Using these equations and the refractive index data for deuterated Potassium Dihydrogen Phosphate (KD*P) (Yariv, 1976), illuminated with 413nm light at this angle, the downconverted light at 826nm is emitted into a cone of half angle 9.8°. This is internal to the crystal, and corresponds to 14.3° externally. A broad spectrum of light is emitted at angles around this.

Two near identical photon streams can be selected from the cone by placing apertures at opposite ends of the cone base, at angles satisfying the phase matching conditions (equations 2-4). A more detailed discussion of the coincidence properties of the photon pairs will be given in section 4.2

2.2 Antibunching experiments

Given two identical trains of photons, one can use the detection of a photon in one train (trigger channel) to gate detection of photons in the other train. Putting a dead time between gate open events can, in principle, produce an antibunched light source. After each photon there will exist a reduced probability of seeing another for a period equal to this dead time. If the photodetection rate in the triggering channel is much higher than the inverse dead time (ie we overdrive the system) the gate open events will become regularly spaced. This leads to a reduction in the noise of the gated beam to below that of Poisson light.

The experiment can be more easily understood by referring to the schematic diagram shown in figure 2. It is a modified version of the original apparatus used by Walker and Jakeman (1984; 1985) where the shutter was placed before the non-linear crystal. Splitting of incident UV photons takes place within the non-linear crystal and the pairs of angularly resolved photons so produced are observed by (photon counting) trigger and signal detectors. At the trigger detector an electronic dead time τ_D is introduced so that detection events in this channel are at least τ_D apart. When each of these events are registered an optical shutter in the other, signal channel, is opened for a short time t. An optical delay is included before the shutter to compensate for electronic and shutter response times, etc. Thus only partners of events actually registered in the trigger channel should be detected in the signal channel. In principle this overcomes the problem of random partnerless events arising in the signal channel due to the low efficiency of the trigger channel detector, which considerably reduced the the non-classical effects observed in the earlier experiment (Walker and Jakeman, 1985; 1986). There is, in practice, always some chance of registering further events in the signal channel during the finite open time t.

Given N shutter openings per sample time T with $t \ll T, \tau_D$ and assuming initially that the signal channel detector is 100% efficient we can write the number of counts in the signal channel in a typical sample time as

$$n = N(1+m) \tag{5}$$

where m is drawn from a Poisson distribution of mean $(1-\eta_1)rt$. Here r is photon pair arrival rate without losses and η_1 is the trigger detector efficiency. Using equation 5 we can express the normalised second factorial moment of the photocount n in the form

$$\begin{aligned} n^{(2)} &= \frac{<n(n-1)>}{<n>^2} \\ &= \frac{<N(N-1)>}{<N>^2} \times \left[1 + \frac{<m>}{(1+<m>)^2}\right] + \frac{<m>(2+<m>)}{<N>(1+<m>)^2} \end{aligned} \tag{6}$$

angular brackets denoting ensemble averages with sample time T. This quantity is not affected by the finite quantum efficiency of the signal detector. The reduction in measured signal variance over the conventional Poisson variance,the Fano factor F , is however dependent on signal channel quantum efficiency. By definition

$$F = (\text{Var } n)/<n>= 1+<n>(n^{(2)}-1) \tag{7}$$

where $<n>= \eta_2 <N>(1+<m>)$ and $n^{(2)}$ is given by eq. 7. As photon counting detector efficiencies are low we do not expect a large reduction in Fano factor in this experiment. On the other hand the second factorial moment is minimised by choosing $T < \tau_D$ when the first term in eq. 6 vanishes. From the theory of deadtimes (Muller, 1974)

$$<N>= \frac{\eta_1 rT}{1+\eta_1 rT} \tag{8}$$

we can write in this case (Jakeman and Jefferson, 1986)

$$n^{(2)} = \frac{1+\eta_1 r\tau_D}{\eta_1 rT}\left(1 - \frac{1}{1+(1-\eta_1)^2 rt}\right) \tag{9}$$

The first term in eq. 9 is a bunching term as $T \leq \tau_D$ and is reduced by overdriving the trigger channel ($\eta_1 r\tau_D \geq 1$). However the second term shows an antibunching effect when η_1 is large and rt the number of photon pairs arriving in the gate time is small. Hence a short gate time and high trigger channel detection efficiency are desirable. It must also be noted that any losses or misalignment of the phase matched apertures will reduce the effective efficiencies η_1 and η_2 appearing in eqs. 7 and 9

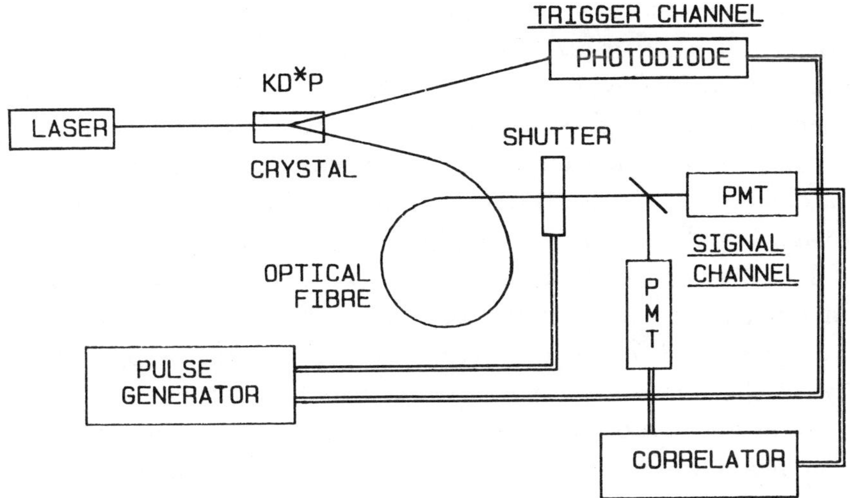

Figure 2: Block diagram of the photon antibunching apparatus.

These effective quantum efficiencies can be estimated by measuring the photodetection event coincidences between trigger and signal channels $\overline{C}$

$$\overline{C} = \eta_1 \eta_2 r \tag{10}$$

and dividing by the mean count rate in each channel

$$\overline{n_j} = \eta_j r \qquad j = 1, 2 \tag{11}$$

To avoid detector afterpulsing effects one can estimate $n^{(2)}$ from the cross-correlation function $g_x^{(2)}(\tau)$ from two detectors viewing the signal channel via a beamsplitter. It has been shown that

$$g_x^{(2)}(\tau) = \frac{< n_1(0) n_2(\tau) >}{< n_1 >< n_2 >} = \begin{cases} g^{(2)}(\tau) & \tau \neq 0 \\ n^{(2)} & \tau = 0 \end{cases} \tag{12}$$

We expect a positive zero time slope in $g_x^{(2)}$ when the light is antibunched and this initial dip in the correlation function is below unity when the light is also sub-poissonian. In an initial experiment where the effect of the shutter was simulated by postdetection electronic gating quite large effects were seen (Brown et al, 1986a)

Here we discuss in more detail the all optical experiment (Rarity et al, 1987a). A schematic of the apparatus is shown in figure 2. A KD*P crystal illuminated by a 325nm Helium/Cadmium laser was used; tilted until the red ($\sim$ 650nm) downconverted photons subtended an an-

about 10°. For maximum η_1 an avalanche photodiode operated in 'geiger' mode (Brown et al, 1986b; 1987) was used as trigger channel detector. Operated at 25 volts beyond breakdown and passively quenched using a 330KΩ series resistor this device shows 25% quantum efficiency at −3°C.

The optical delay consisted of 170 m of multimode optical fibre producing a delay of 0.9μs. At the output of the fibre a X15 magnification system transferred an image of the 50 μm core to a fast acousto-optical switch. The first order Bragg spot could be switched on after a delay of about 0.6 μs with a rise time less than 120 ns. Using the first order beam an extinction ratio better than 250:1 could be obtained but only 30% of the incident light was transmitted. Improvement in transmission could only be obtained at the expense of switching speed. The first order beam was then imaged onto the photocathodes of two conventional photon counting photomultiplier (PMT) systems via a pellicle beamsplitter.

Conjugate photon trains were selected by fibre input and trigger detector apertures combined with long wavelength transmitting colour glass filters (cut-off below 600 nm), to remove UV flare and fluorescence. Aperture positions were optimised by maximising the coincidence count rate between the two channels (PMT versus APD) with the shutter held open. This operation led to effective channel efficiencies of $\eta_1 = .09$ and $\eta_2 = .0006$ being measured using eqs. 10 and 11. This suggests some fluorescence dilution or alignment losses as the trigger channel theoretical efficiency is .25. Other losses arise in the signal channel due to the fibre and shutter (transmission 0.1) and an effective photomultiplier quantum efficiency of only .01. As sample time $T \leq \tau_D$, the dead time, we are confined to a mean count per sample time $< n > \leq \eta_2$ hence from eq. 7

$$F \geq 1 - \eta_2 \tag{13}$$

which will be close to unity. However we can achieve a significant antibunching effect in the second order statistics of $n^{(2)}$ within the restrictions of eq. 9.

Two typical experimental results are shown in fig. 3 along with a control result where crystal and laser were replaced by a simple tungsten filament light source attenuated to similar brightness. Clearly both results show an antibunching and sub-poissonian effect with $g^{(2)}(0) < 1$. For result 1 the shutter-open time was made as short as possible ($t \sim 150ns$) with loss of efficiency. This allowed a dead time $\tau_D = 20\mu$s and

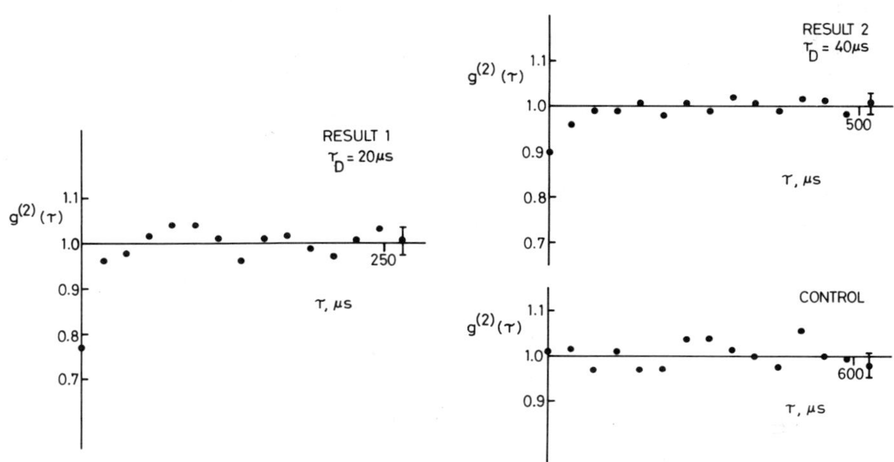

Figure 3: Normalised correlation function $g^{(2)}$ of signal channel counts as a function of delay time τ.

sample time $T = 19\mu s$ to be used resulting in a strong antibunching with $g^{(2)}(0) = .77$. Dark count correction leads to an estimated $n^{(2)} \sim 0.42$ but the low count rate limits the Fano factor to $F = .99980 \pm 3 \times 10^{-5}$. This is still only a 2% effect predetection (assuming 1% effective detection efficiency). Result 2 used a longer shutter open time of $.5\mu s$ with consequent reduction of the anti-bunching effect (larger rt factor in eq. 9) but the higher transmission leads to a similar Fano factor for the detected light.

Table 1.Summary of antibunching results[a]

Result	t (μs)	T (μs)	τ_D (μs)	$\overline{n}$ (cps)	$n^{(2)} \equiv g^{(2)}$ [b]	Fano factor
1	.15	19	20	44.9	0.77 (.42) ±.03	.99980 $\pm 3.10^{-5}$
2	.5	37	40	58.8	.905 (.83) ±.017	.99983 $\pm 4.10^{-5}$
Control	.5	45	50	60.3	1.007 (1.004) ±.02	1.0

[a](taken from Rarity et al, 1987a)
[b](figures in brackets are dark count corrected $n^{(2)}$)

A summary of the results is shown in table 1. As the count rates wer so low the results had to be averaged over long durations to sho nificant effects. Although these results are clearly sub-poissonian the light source is not useful for reduced noise measurements. Further reduction in Fano factor in this configuration can only be achieved by reducing the signal channel losses. This could be done by using a shorter length delay with a faster responding shutter (eg an electro-optic modulator) and higher efficiency solid state photon counting detectors of a type similar to that used in the trigger channel (Brown et al, 1986b; 1987). However, any optics introduced after the crystal will introduce some losses which in turn will reduce the noise reduction achievable. Furthermore the power output from this type of device is fundamentally limited to count rates less than the inverse shutter-open time t^{-1}. As an alternative to this feedforward technique one could return to the original feedback technique used by Walker and Jakeman (1985). Here an open shutter was placed before the crystal and closed for a dead time after each trigger channel photodetection. Dilution of the non-classical effects by unpaired photons would not be large if trigger channel efficiency were near unity but again the power would be limited by the response time of the feedback loop and shutter. The alternative to these single photon techniques are band-limited feedback systems using analogue detection where high detector efficiencies can be exploited.

2.3 Sub-poissonian light produced using analogue feedback

For this work we use an arrangement (figure 4) which is similar to that of Walker and Jakeman's (1984; 1985) first photon counting experiments. However, in this experiment the *analogue* shot noise from one photon stream is negatively fed back to the modulator. This greatly reduces the noise in the closed loop, and also reduces the noise from the other, correlated photon stream.

This type of feedback system can be studied using a simple phenomenological theory (Tapster et al 1988). Before introducing any feedback the photocurrents (rather than count rates) from each detector are given by

$$i_1 = \eta_1 r e, \qquad i_2 = \eta_2 r e \tag{14}$$

where e is the electron charge. Associated with each photocurrent is a shot noise Δi given by the well known shot noise formula

$$< \Delta i_j^2 >= 2 e i_j \Delta f \qquad j = 1, 2 \tag{15}$$

A fraction $\eta_1\eta_2$ of the photon pairs are seen by both detectors. This corresponds to a (coincidence) photocurrent

$$i_3 = \eta_1\eta_2 re \tag{16}$$

The remainder of the noise currents are uncorrelated, therefore their cross spectrum is the shot noise in this coincidence current

$$< \Delta i_1 \Delta i_2 >= 2ei_3\Delta f \tag{17}$$

We can define a normalised cross spectrum as

$$s_{12} = \frac{< \Delta i_1 \Delta i_2 >}{\sqrt{< \Delta i_1^2 >< \Delta i_2^2 >}} = \sqrt{\eta_1\eta_2} \tag{18}$$

on combining equations 15,16 and 17. The shot noise currents and their cross spectrum can be measured directly when the feedback loop is disconnected.

If a (small) fraction k of the normalised photocurrent fluctuations $\Delta I_2/i_2$ from detector 2 is negatively fed back to the modulator fluctuations in channel 2 are reduced. To first order

$$\Delta I_2 = \Delta i_2/(1+k) \tag{19}$$

where ΔI_2 is the new photocurrent fluctuation.The fluctuations in channel 1, the open loop channel, then become

$$\Delta I_1 = \Delta i_1 - \frac{i_1 k \Delta i_2}{i_2(1+k)} \tag{20}$$

Using eqs. 13,14,15 and 16 the mean square fluctuations in channel 1 are

$$< \Delta I_1^2 >= (2ei_1\Delta f)\left(1 - 2\eta_1\frac{k}{1+k} + \frac{\eta_1}{\eta_2}\frac{k^2}{(1+k)^2}\right) \tag{21}$$

The optimum value of k minimises eq. 21. Equating the differential with respect to k to zero, and solving gives

$$k = \eta_2/(1-\eta_2) \tag{22}$$

Substituting for k in equation 20 leads to a Fano factor

$$F = \frac{< \Delta I_1^2 >}{(2ei_1\Delta f)} = 1 - \eta_1\eta_2 \tag{23}$$

which is less than one for all efficiencies. Substituting equation 21 in 18, squaring and averaging, gives the closed loop mean square fluctuations in channel 2

$$< \Delta I_2^2 >= (2ei_2\Delta f)(1-\eta_2)^2 \tag{24}$$

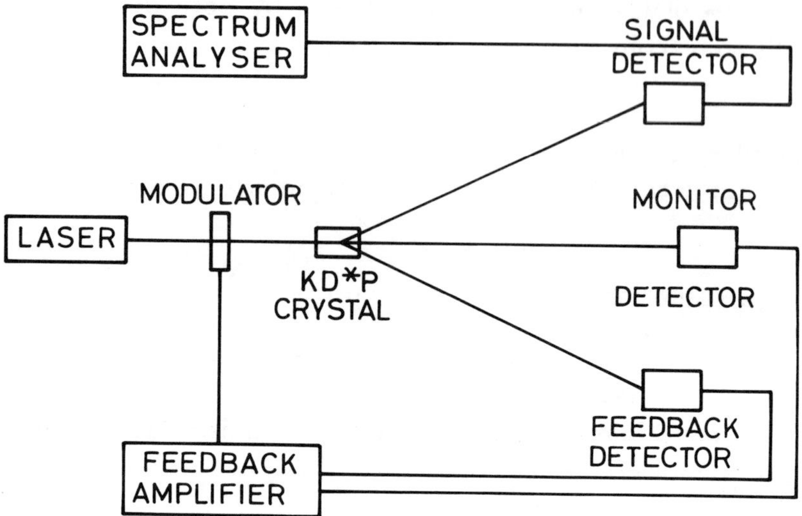

Figure 4: Block diagram of the analogue feedback experiment.

By increasing the feedback gain k, the noise in the feedback channel can be reduced to zero. However, away from the optimum feedback gain, the noise in the other channel increases. In the limit $k \to \infty$, the channel 1 Fano factor becomes

$$F = 1 - \eta_1(2 - 1/\eta_2) \tag{25}$$

This is only less than 1 for $\eta_2 > .5$.

Noise due to dark current and amplifier noise can be incorporated in this theory by defining effective efficiencies η'_1, η'_2 given by

$$\eta'_j = \eta_j/(1 + d_j/i_j) \qquad j = 1, 2 \tag{26}$$

where d_j is the dark current which would generate an equivalent excess noise in the detectors.

If the apertures are not ideally aligned then some fraction of the light passing through each aperture will be uncorrelated. This is clearly equivalent to a reduction in the efficiency of the detectors, as measured by the normalised cross spectrum. Losses in the optical system will have a similar effect. The results so far have been independent of frequency. In practice the detectors and amplifiers have a limited bandwidth, and amplifier noise is also frequency dependent. This can be taken into account by making the feedback gain k and effective efficiency η frequency dependent. The feedback gain should also be complex. These factors determine the frequency range over which significant results are achieved.

In the experiment (figure 4) a Krypton ion laser producing 1W of light at 413.4nm passes through an electro-optic modulator (EOM) then is weakly focused into the KD*P crystal. Most of the beam is transmitted through the crystal to a monitor detector and the difference between this signal and a reference is fed back to the EOM to suppress classical fluctuations in the laser beam.

A pair of apertures are placed so as to select matched pairs of photons, and the light is focused onto two detectors. Colour glass filters (<700 nm cut-off) are again used to minimise the amount of scattered laser light and fluorescence reaching the detectors. The detectors are conventional silicon PIN photodiodes (BPX65), cooled to $\sim -20°$C and operated with a reverse bias of 5V. They are connected to low noise (<1fA $\mathrm{Hz}^{-\frac{1}{2}}$ input noise) transimpedance amplifiers with -3dB bandwidths of 350Hz and 1kHz. The optimum noise performance occurs in the range 5 to 100Hz. The 1kHz bandwidth detector can be connected via a variable gain AC coupled amplifier to the reference input of the feedback amplifier. This has the effect of modulating the laser beam with the detector output. The other detector is connected via an AC coupled low pass filter to a computer performing as an averaging FFT spectrum analyser. The computer could also perform spectral cross correlation of the signals from the two detectors. The mean values of the photocurrents were measured at the outputs of the detector amplifiers, using calibrated amplifier gains.

The apertures subtend about 40 milliradians at the crystal selecting downconverted wavelengths of 830$\pm$60nm with detected photocurrent $\sim$30pA. This corresponds to $\sim$60pW of light and 2.10^8 photons per second. With a bandwidth of 1kHz the normalised shot noise $\Delta i/i$ has rms value $\sim 3.10^{-3}$. The alignment of the apertures was optimised by maximising the normalised cross spectrum of the two detector outputs (eq. 18). The cross spectrum after alignment is shown in figure 5, and the average correlation coefficient in the 30 - 90Hz region is 0.48. The theory above implies that with a correlation coefficient of $\sqrt{\eta_1'\eta_2'} = 0.48$ (eq. 18) a Fano factor of 0.77 is expected (eq. 23), when the optimum feedback is used. Assuming that both effective efficiencies are the same, the feedback gain was chosen to reduce the noise in the feedback detector to $(1-0.48)^2$ (eq. 24) of its original value.

A typical set of results is shown in figure 6. The average noise power in the range 30 to 90Hz, excluding the point at 50Hz is just 0.78 of that for the calibration spectrum which agrees well with the expected

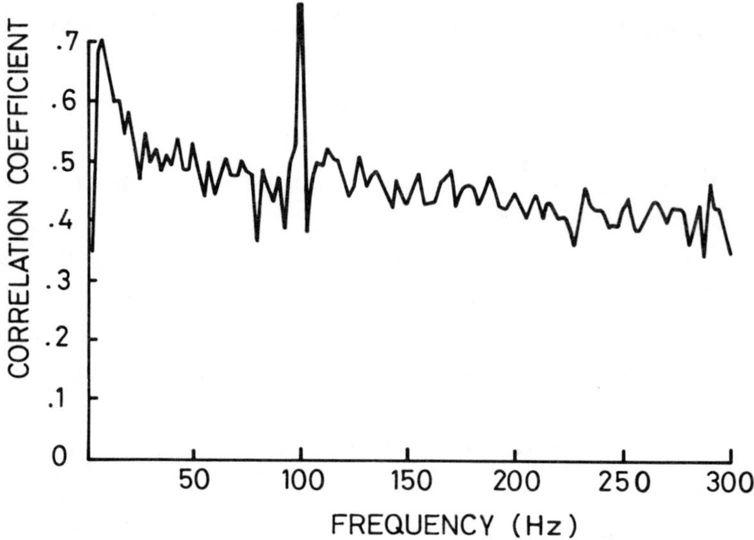

Figure 5: The normalised cross-spectrum s_{12} of the photocurrent fluctuations from the two detectors. If the detectors have identical performances this is the effective quantum efficiency of each detector.

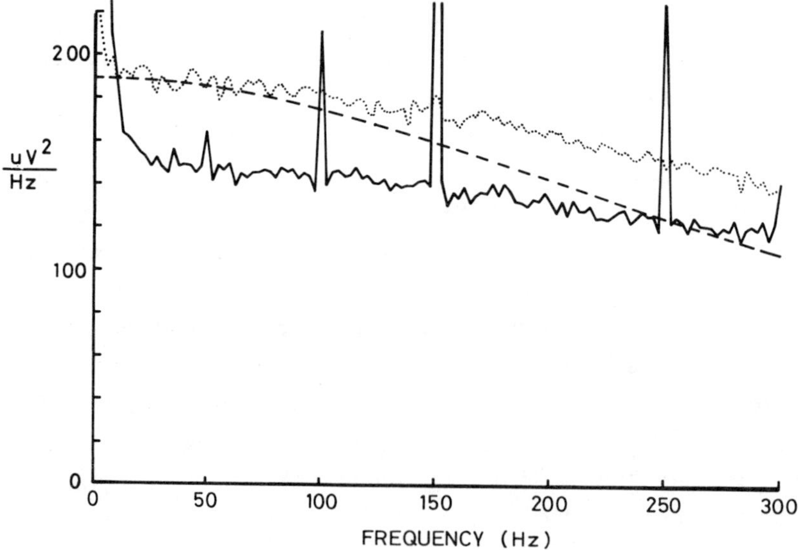

Figure 6: The power spectrum of the photodetector output when detecting sub-poissonian light (full curve), compared with the calibration spectrum of Fano factor unity (dotted curve).The theoretical curve (dashed line) is calculated using the measured amplifier bandwidth, the detector DC transresistance of 4.67GΩ and measured DC photocurrent of 26.6 pA.

result. Assuming the detector quantum efficiency is about 80% at the wavelengths used, and allowing for the dark noise, we infer a predetection Fano factor of 0.72. The dashed calibration spectrum shows a theoretical shot noise spectrum calculated using the calibrated amplifier gains and the measured mean photocurrent. The dotted curve shows the spectrum from an equal power shot noise source (a highly attenuated light emitting diode Tapster et al, 1987). The difference between the two calibration curves gives an indication of the amplifier noise contribution.

If detector efficiency were the only limiting factor one would expect to measure a correlation coefficient in the region .75 to .8. There are thus other losses which could be reduced to improve on this result, principally: optical surface reflection losses, fluorescence dilution of the photon pairs, non-optimal aperture alignment and detector dark noise. Choosing a crystal with a higher optical non-linearity would reduce the importance of fluorescence and detector noise while creating a higher power source. Higher quantum efficiency detectors would clearly also improve the results. Internal quantum efficiencies of silicon detectors are above 95% for wavelengths below 830nm (Korde and Geist, 1987). Higher power sources could also be achieved by using type II degenerate phase matching within a cavity. Heideman et al (1987) report sub-poissonian intensity differences from such a twin beam source (see also Giacobino and Fabre, 1988; in this volume).

3 USES FOR NON-CLASSICAL LIGHT SOURCES

3.1 Quantum efficiency measurement

The principle of using pair sources for measurement of absolute source activity and detector quantum efficiency is well known in the field of nuclear physics (Knoll 1979). Here we are only extending this principle to the, lower energy, optical region. Returning to the equations for the coincidence count rate $\overline{C}$ and photodetection rates in the photon counting experiments (eqs. 9 and 10)

$$\overline{C} = \eta_1 \eta_2 r \tag{27}$$

$$\overline{n_j} = \eta_j r \qquad j = 1,2 \tag{28}$$

we see that the channel quantum efficiency can be estimated by dividing the coincidence count rate by the mean count rate in the other channel.

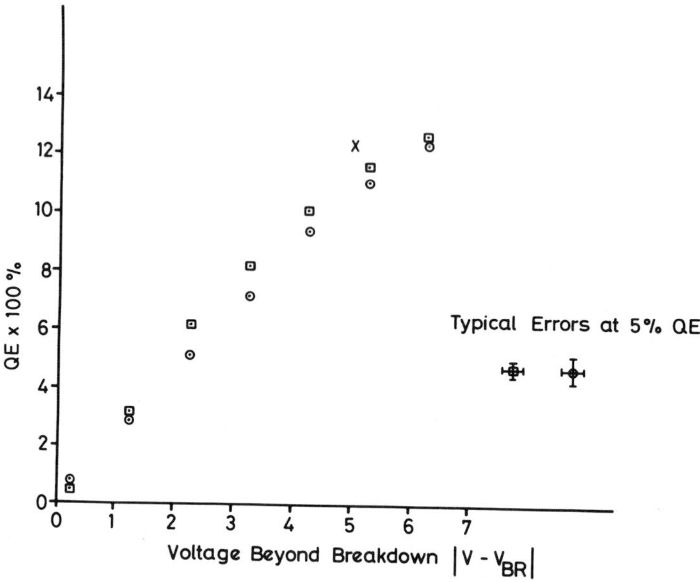

Figure 7: Avalanche photodiode quantum efficiency measured as a function of voltage beyond breakdown ($V - V_{BR}$). □ determined using the coincidence method described here, ⊙ determined using calibrated neutral density filters and a source of known brightness. X estimated from manufacturers data sheet.

By ensuring that losses in one channel are small one can make an absolute measurement of detector quantum efficiency in that channel. This can be done by ensuring that all frequency selection and aperturing are carried out on the conjugate beam. A typical result of such a measurement (Rarity et al, 1987b) is shown in figure 7 where the quantum efficiency of an actively quenched APD photon counting detector (Brown et al, 1987) is plotted as a function of bias voltage (beyond breakdown). An apparatus similar to that shown in figure 2 was used. 650±5nm photons were selected by placing an interference filter in the fibre channel with only a wide aperture and fluorescence filter (transmission 93%) limiting the light reaching the APD detector. The comparison conventional measurement was made using a helium/neon laser (633nm) attenuated by $\sim 10^{-9}$ using calibrated neutral density filters. The unattenuated laser output was measured using a calibrated photodiode power meter.The coincidence data fall slightly above the laser results due to the increase in device quantum efficiency between 633 and 650nm.

The relative standard deviation σ_R of the measurement as a function of total coincidence counts N_c can be shown to be

$$\sigma_R = \left[\frac{1}{N_c}(1-\eta)\right]^{1/2} \tag{29}$$

This result is sub-poissonian and can become arbitrarily accurate as η approaches unity.

In an analogue apparatus quantum efficiency can also be measured from the noise correlation (Sergienko and Penin, 1987). Adapting eq. 17 we obtain

$$\eta_1 = \frac{< \Delta i_1 \Delta i_2 >}{< \Delta i_2^2 >} \tag{30}$$

The accuracy of such a measurement will be limited by the excess noise in the detector amplifier and correlations from macroscopic laser fluctuations occurring in the measurement bandwidth.

The broadband nature of the downconverted cone means that in principle detector quantum efficiencies can be measured over a wide bandwidth using a variable monochromator and linked goniometer to preserve the phase matching angles.

3.2 Reduced quantum noise in transmission measurements

Changes in the intensity due to scattering,absorption,reflection or refraction by a sample are fundamental to microscopy and imaging techniques generally and the absorption spectrometer is a standard piece of laboratory equipment finding uses in many scientific disciplines. In the case, often occurring in bio-medical measurements, where the specimen can be damaged or optically changed on illumination there is an upper limit to the intensity and exposure time. Under this dose limit or power/bandwidth constraint (Yuen 1988) the accuracy of the measurement can be improved by reducing the number uncertainty in the incident beam. However the loss in the sample itself will limit the noise reduction achievable.

In a conventional transmission measurement only one channel is used. The mean flux of photons from the source r (or the mean detected flux $r' = \eta r$) can be measured to arbitrary accuracy before introducing the sample into the apparatus. Introducing a sample with absorption coefficient $0 \le \alpha \le 1$ into the beam for a finite time T and counting (or integrating the photocurrent in an analogue apparatus) the number s of transmitted photons with a detector of quantum efficiency η we can estimated α from

$$\hat{\alpha} = 1 - s/r'T \tag{31}$$

with measurement variance

$$\text{var}(\hat{\alpha}) = <\alpha^2> - <\alpha>^2 = \frac{1-\alpha}{r'T} \tag{32}$$

given that s is a Poisson random variable.

A measurement made using a correlated pair source such as our parametric downconversion source (without feedback) would involve placing the sample in one channel of the apparatus and measuring direct, n and transmitted, s counts in sample time T. An estimate of α can be obtained from

$$\hat{\alpha} = \frac{n-s}{n} \tag{33}$$

where for simplicity we assume equal detector quantum efficiencies η. n and s are Poisson random variables with correlated noise

$$<ns> - <n><s> = <\Delta n \Delta s> = \eta^2(1-\alpha) \tag{34}$$

hence to order $<n>^{-2}$ (Jakeman and Rarity, 1986),

$$\text{var}(\hat{\alpha}) = \frac{(1-(2\eta-1)(1-\alpha))(1-\alpha)}{r'T} \tag{35}$$

This estimator is slightly biased but the bias is always less than the error of measurement. We can define a variance ratio (or Fano factor) relating the conventional (C) to the correlated pair (P) measurement as

$$R = \frac{\text{var}(\hat{\alpha_P})}{\text{var}(\hat{\alpha_C})} = 1 - (2\eta - 1)(1-\alpha) \tag{36}$$

There is no reduction in variance possible until η is greater than 0.5. Using an estimator based on partial use of the pair count

$$\hat{\alpha} = \frac{(1-k)<n> + kn - s}{(1-k)<n> + kn} \tag{37}$$

and choosing $k \sim \eta(1-\alpha)$ one can obtain a result which is always sub-poissonian with

$$R = 1 - \eta^2(1-\alpha) \tag{38}$$

In high sensitivity spectroscopy techniques where the absorption is extremely small the stability of the source at low frequency can be the limit to sensitivity. Frequency modulation techniques are thus used to move the signal out of the noise dominated range and differential intensity measurements (across a spectral feature) are made which can, with care, lead to shot noise limited measurement of absorption (Wong and Hall, 1985; Gehrtz et al 1985). In these situations both the attenuated and unattenuated beams have to be estimated from a limited time measurement and the estimator for α is similar to eq. 33. However in a conventional apparatus the correlation (eq. 34) is zero and

$$\mathrm{var}(\hat{\alpha}) = \frac{2-\alpha}{r'T} \tag{39}$$

When compared with the photon pair subtractive estimator (eq. 35) this leads to an improvement for all detector efficiencies with

$$R = 1 - \frac{2\eta(1-\alpha)}{(2-\alpha)} \tag{40}$$

When α is small this result becomes

$$R \sim 1 - \eta \tag{41}$$

which is equivalent to the result of Giacobino and Fabre (1988). Using our apparatus as a source of correlated photons in a differential absorption measurement of this type would reduce the variance by a factor of 2 below the shot noise level.

In the general case of sample illumination using a sub-poissonian source, either the parametric source described above or a direct source such as a high efficiency light emitting diode (Tapster et al, 1987) or semiconductor laser (Machida et al, 1987), the reduction factor is given by

$$R = 1 - \eta(1-\alpha)(1-F) \tag{42}$$

where F is the source predetection Fano factor.

Other estimators can be based on coincidence measurements (Jakeman and Rarity, 1986) as in the detector quantum efficiency measurement discussed above. There the detector can be thought of as an ideal unit efficiency device behind an absorber with $\alpha = 1 - \eta$.

In all absorption measurements the noise reduction is always limited by

$$R \geq \alpha \tag{43}$$

hence the technique is limited to highly transmissive samples and all optical losses (eg container windows, filters etc.) have to be minimised.

4 PHOTON LOCALISATION AT A BEAMSPLITTER

4.1 Introduction

Several simple quantum mechanical models of beamsplitters have recently been published (Fearn and Loudon, 1987;Ou et al, 1987; Prasad et al, 1987). Such a beamsplitter appears to interact with single beams of classical and non-classical light in the same way as macroscopic models. One simple result is the binomial partition noise imparted on an N-photon state due simply to the random selection of output direction for each input photon (Brendel et al,1988). The beamsplitter is, however, a four port device and a less intuitive result is obtained when indistinguishable N-states appear simultaneously at both input ports of the beamsplitter. The simplest case of this type occurs when one-photon states are simultaneously input into each port of a 50/50 beamsplitter (Fearn and Loudon, 1987). The theory predicts that each output will contain either two photons or none. A naive (particle) theory would predict a mixture of two's and one's. The required $\mid 1,1 >$ photon state can be selected in non-degenerate parametric downconversion experiments by photon counting coincidence detection (Mollow, 1973). The disappearance of this coincidence after a beamsplitter would be expected if the theory is correct. In a more rigorous theory this effect will occur over a range of path length differences due to the finite length of the photons which can be related, by fourier transform, to their bandwidth. This in turn can be related to the geometry of the parametric downconversion apparatus and the detailed phase matching conditions at the crystal.

We have constructed a white light interferometer (section 4.4) behind a downconversion crystal which ensures that photons from the same spatial mode, and within the same bandwidth arrive at the two inputs of a beamsplitter. We demonstrate that the effect only occurs if the difference of the arrival times is within the coherence time or inverse bandwidth of the downconverted photons. As these photons have a bandwidth of a few nanometres, this time is short (less than 100fs).

4.2 Coincidence properties of parametric photon pairs

The two photon detection probability P_{12}, that of detecting photons at times t_1, t_2 at points $\vec{X}_1$, $\vec{X}_2$ from a source centred at the origin can be obtained from

$$P_{12} \propto < \hat{A}_1^{(-)}(t_1, \vec{X}_1)\hat{A}_2^{(-)}(t_2, \vec{X}_2)\hat{A}_2^{(+)}(t_2, \vec{X}_2)\hat{A}_1^{(+)}(t_1, \vec{X}_1) > \tag{44}$$

where $\hat{A}^{(+)}(t)$ is the positive frequency part of the electric field operator (which can also be described as a photon annihilation operator) at time t. For a spontaneous parametric downconversion source (or in the general low flux limit) Mollow showed that

$$P_{12} \propto | < \hat{A}_1^{(+)}(t_1, \vec{X}_1)\hat{A}_2^{(+)}(t_2, \vec{X}_2) > |^2 \tag{45}$$

is a good approximation. The product $\hat{A}_1^{(+)}(t_1, \vec{X}_1)\hat{A}_2^{(+)}(t_2, \vec{X}_2)$ can be thought of as a two photon wavepacket. For brevity in the ensuing discussion we define the shorthand

$$g_{12} \equiv < \hat{A}_1^{(+)}(t_1, \vec{X}_1)\hat{A}_2^{(+)}(t_2, \vec{X}_2) > \tag{46}$$

For a finite size crystal evaluation of g_{12} involves a coherent summation over the illuminated region of all the conjugate frequencies contributing to the two photon wavepacket (Mollow, 1973)

$$\begin{aligned} g_{12} \propto & \int\int d\omega_1 d\omega_2 \omega_1 \omega_2 \chi^{(2)}(\omega_0, \omega_1, \omega_2)\delta(\omega_0 - \omega_1 - \omega_2) \\ & \times \int d^3r e^{-i[\omega_1(t_1 - X_1'/c) + \omega_2(t_2 - X_2'/c)]} . e^{-i\vec{k}_0 . \vec{r}} f(\vec{r}) \end{aligned} \tag{47}$$

where $\chi^{(2)}$ is the relevant component of the second order non-linear optic tensor, $\omega_0, \vec{k}_0$ are the angular frequency and wavevector of the incident beam, $\omega_{1,2}$ the angular frequencies in the downconverted photons and c is the speed of light in vacuo. $f(\vec{r})$ defines the illuminated region within the crystal as a function of position $\vec{r}$ measured from the centre of the crystal. The convention of an incident plane wave beam travelling in the z direction is assumed. The derivation of $X'_{1,2}$ for a crystal of refractive index $n(\omega_{1,2})$ is illustrated in figure 8 where we see

$$X'_j = | \vec{X}_j - \vec{r}' | \qquad j = 1, 2 \tag{48}$$

where $\vec{X}_j$ is now the vector connecting the 'virtual' crystal centre to the aperture j (ie has phase length X_j/c measured from the crystal centre) and

$$\vec{r}' = (r_x, r_y, r_z/n) \tag{49}$$

under the apparent depth approximation valid at low angle θ. Use of the cosine rule to evaluate X'_j and assuming $r/X \ll 1$ for all the illuminated

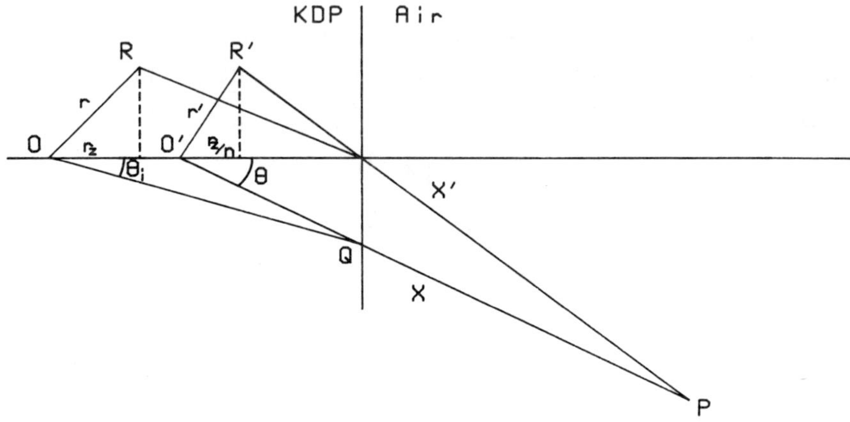

Figure 8: The geometry relating $\vec{X}$, $\vec{X}\,'$, $\vec{r}$ and $\vec{r}\,'$

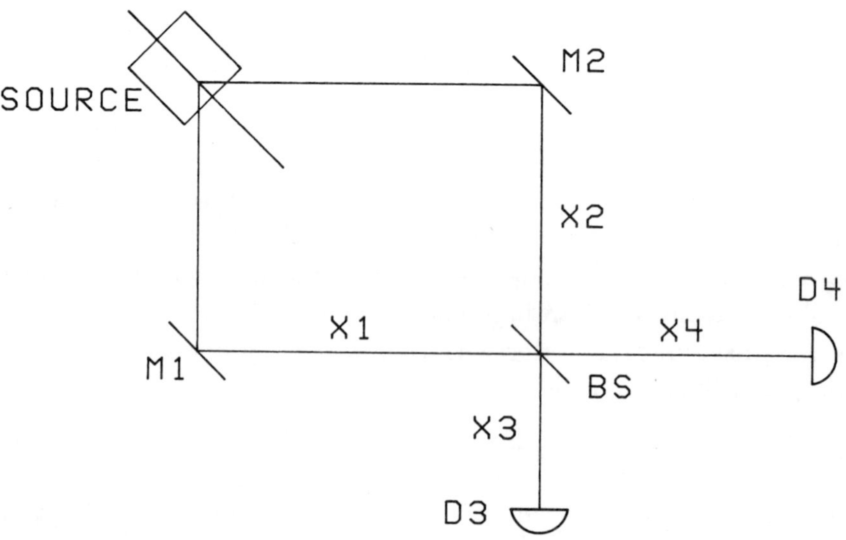

Figure 9: Schematic of the beamsplitter BS labelling inputs 1,2 and outputs 3,4 and related distances from a source X_1, X_2 and to detectors X_3, X_4.

region leads to

$$g_{12} \propto \int\int d\omega_1 d\omega_2 \omega_1 \omega_2 \chi^{(2)}(\omega_0, \omega_1, \omega_2)\delta(\omega_0 - \omega_1 - \omega_2) e^{-i[\omega_1 \bar{t}_1 + \omega_2 \bar{t}_2]}$$
$$\times \int d^3r \; e^{i(\vec{k}_1.\vec{r}_1' + \vec{k}_2.\vec{r}_2' - \vec{k}_0.\vec{r}_0')} f(\vec{r}) \tag{50}$$

where for brevity we use reduced time $\bar{t} = t - x/c$ and $\vec{k}_j = \omega_j \vec{X}_j / cX_j$ and imply the frequency dependence of the refractive index in the r_j . The space integration now clearly leads to the simplistic phase matching criterion described in eq. 1. Specialisation to the symmetric case where external angles are equal $\theta_1 = \theta_2 = \theta$ and where the exact phase matching (eqs. 1,2) occurs at the centre of the crystal we obtain

$$g_{12} \propto e^{-i\omega_0(\bar{t}_1 + \bar{t}_2)/2} \int d\omega(\tfrac{\omega_0}{2} + \omega)(\tfrac{\omega_0}{2} - \omega)\chi^{(2)}(\omega_0, \omega) e^{-i\omega(\bar{t}_1 - \bar{t}_2)}$$
$$\int d^3r \; e^{i[2r_x\omega \sin\theta/c]} e^{-i[2r_z.\omega^2 D \cos\theta/cn_o^2]} f(\vec{r}) \tag{51}$$

where

$$\omega_1 = \frac{\omega_0}{2} + \omega \qquad\qquad \omega_2 = \frac{\omega_0}{2} - \omega \tag{52}$$

satisfying the energy conservation delta function. We also include a dispersion D defined by $n(\omega) = n_o(1 \pm \omega D/n_o)$ with n_o the ordinary refractive index at the symmetric frequency. Assuming a beam with $1/e$ intensity width σ and incident on a crystal of length L we obtain

$$g_{12} \propto e^{-i\omega_0/2(\bar{t}_1 + \bar{t}_2)} \int d\omega(\tfrac{\omega_0}{2} + \omega)(\tfrac{\omega_0}{2} - \omega)\chi^{(2)}(\omega_0, \omega) e^{-i\omega\delta t}$$
$$\times e^{-\omega^2\sigma^2 \sin^2\theta/c^2} \mathrm{sinc}(\omega^2 DL\cos\theta/2cn_o^2) \tag{53}$$

where we define $\delta t = \bar{t}_1 - \bar{t}_2$.

4.3 Quantum theory of the lossless beamsplitter

The beamsplitter is a four port device with input ports 1,2 and output ports 3,4 as shown in figure 9. Coincidence detection rates will be proportional to P_{34} as defined in eq. 44 with suitable change of subscripts. Following Fearn and Loudon (1987) we can define the beamsplitter in terms of a complex amplitude transmission coefficient t_b and reflection coefficient r_b satisfying

$$| r_b |^2 + | t_b |^2 = 1 \qquad\qquad t_b r_b^* + r_b t_b^* = 0 \tag{54}$$

where to connect with Hong and Mandel (1987) $| r_b |^2 = R$ and $| t_b |^2 = T$, the intensity reflection and transmission coefficients. The second part of eq. 54 can be satisfied by assuming a $\pi/2$ phase difference between reflected and transmitted beams. One can express the fields $\hat{A}_{3,4}^{(+)}(t)$ at detectors $D_{3,4}$ placed at distances $X_{3,4}$ beyond the beamsplitter in terms of those at the parametric downconversion crystal $\hat{A}_{1,2}^{(+)}(t)$ on referring to figure 9

$$\begin{aligned} \hat{A}_3^{(+)}(t) &= r_b \hat{A}_1^{(+)}(t - X_T/c + \tau_1 + \delta t) + t_b \hat{A}_2^{(+)}(t - X_T/c + \tau_1) \\ \hat{A}_4^{(+)}(t) &= t_b \hat{A}_1^{(+)}(t - X_T/c + \delta t) + r_b \hat{A}_2^{(+)}(t - X_T/c) \end{aligned} \tag{55}$$

noting that

$$\begin{aligned} X_1 + X_4 &= X_T & X_2 - X_1 &= c\delta t \\ X_2 + X_3 &= X_T - c(\tau_1 + \delta t) & X_4 - X_3 &= c\tau_1 \end{aligned} \tag{56}$$

$c\delta t$ is thus the path length difference between the two arms of the apparatus while τ_1 is a time difference between detections. As the coincidence apparatus has a finite gate time (usually $T \geq 10$ns) we measure a coincidence rate $\overline{C}$ proportional to

$$\overline{C} \propto \int_{-T/2}^{T/2} P_{34}(\tau - \tau_1) d\tau \tag{57}$$

As $T \gg \tau_1, \omega_0^{-1}$ we can, in all practical situations rewrite the limits of the integral as $\pm\infty$ and subsume τ_1 in τ. Forming P_{34} from eqs. 55 and 56 we obtain

$$\overline{C} \propto \left[R^2 + T^2 - RT \left(\frac{\int_{-\infty}^{\infty} g_{12}^*(\tau) g_{12}(2\delta t - \tau) d\tau}{\int_{-\infty}^{\infty} | g_{12}(\tau) |^2 d\tau} + c.c. \right) \right] \tag{58}$$

again assuming $T \gg \delta t$. Evaluation of the τ integral leads to

$$\overline{C} \propto \left[R^2 + T^2 - RT \left(\frac{\int_{-\infty}^{\infty} f^*(\omega) f(-\omega) e^{-i2\omega\delta t} d\omega}{\int_{-\infty}^{\infty} | f(\omega) |^2 d\omega} + c.c. \right) \right] \tag{59}$$

where $f(\omega)$ is a general filter function and it is clear that for large effects there must be a significant overlap between $f(\omega)$ and $f(-\omega)$. For the phase matching limited bandwidth evaluated above

$$f(\omega) = (\frac{\omega_0}{2} + \omega)(\frac{\omega_0}{2} - \omega)\chi^{(2)}(\omega_0, \omega) e^{-\omega^2\sigma^2 \sin^2\theta/c^2} \mathrm{sinc}(\omega^2 \mathrm{DL} \cos\theta / 2\mathrm{cn}_o^2) \tag{60}$$

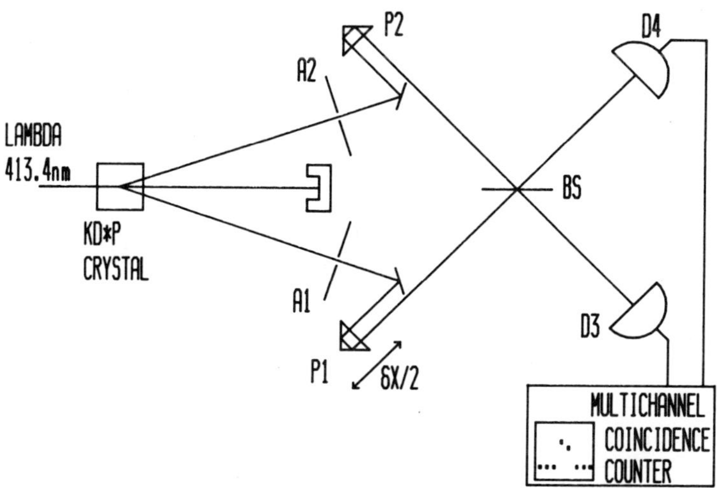

Figure 10: Block diagram of the photon localisation apparatus. A1 and A2 are apertures. Prism P1 is mounted on a micrometer driven translation stage to allow path length difference δx to be varied. The 50/50 beamsplitter combines the two beams which are detected by photon counting detectors D3 and D4.

When dispersion is negligible and the Gaussian term is narrow we can ignore the sinc function, approximate $\omega_0/2 \pm \omega$ by $\omega_0/2$ and assume a constant χ^2. The resulting Gaussian filter function is easily transformed and we obtain

$$\overline{C} \propto R^2 + T^2 - 2RT \exp\left[\frac{-\delta x^2}{2\sigma^2 \sin^2\theta}\right] \tag{61}$$

where $\delta x = c\delta t$.

4.4 Experiment

The experimental arrangement is shown in figure 10. A Krypton ion laser operating at 413.4nm wavelength illuminates a 1cm crystal of KD*P. A pair of small apertures (A1,A2) select matched pairs of photons (826.8nm). The two beams so created are reflected onto a beamsplitter by mirrors and right angle prisms. One of the prisms is mounted on a motorised translation stage allowing the path difference to be varied in 5μm steps. The outputs of the beamsplitter are focused onto two photon counting avalanche photodiode detectors as studied in the quantum

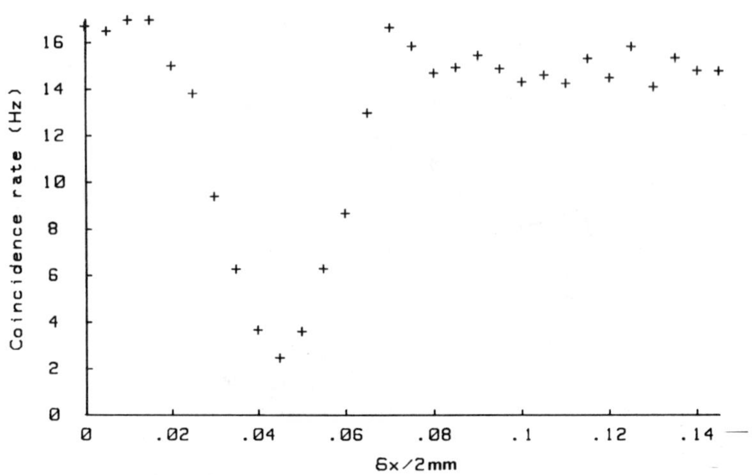

Figure 11: Typical result showing the coincidence rate as a function of prism position with narrow band interference filters in front of detectors.

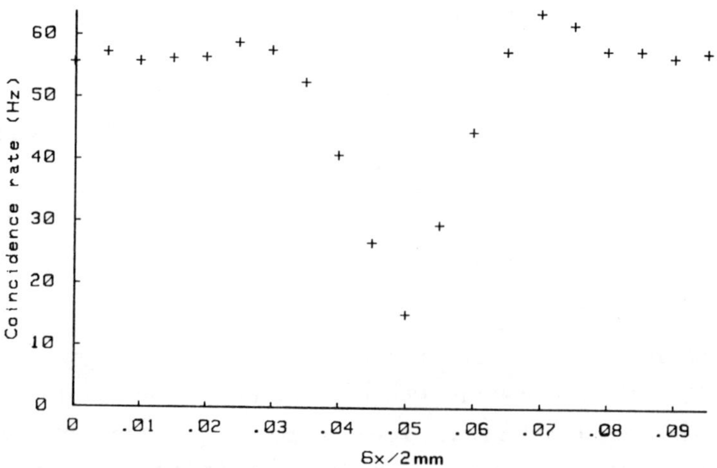

Figure 12: Typical result showing the coincidence rate as a function of prism position with narrow band interference filters removed.

efficiency measurements (section 3.1). The photodetection pulse trains are fed to a single bit correlator operating as a multichannel coincidence counter with 10ns resolution. An HP 300 series computer acquires the correlator data and calculates the two-photon coincidence rate for a series of translation steps through the zero path difference position.

The apparatus was initially set up for maximum coincidence rate without the beamsplitter but with narrowband (± 5 nm) interference filters placed in front of the detectors to select out the symmetric (826.8 nm) phase matching wavelength. The beamsplitter was then introduced and mode overlap was achieved using two helium neon alignment beams crossing at the centre of the crystal then passing through the apertures. The beamsplitter position and alignment was then adjusted to ensure that the recombined beams overlapped both at the beamsplitter and in the far field, when a strongly modulated zero order fringe could be seen.

A typical result (figure 11) taken after this alignment procedure shows the coincidence rate dropping markedly at zero path difference. This result was taken with the 826.8 nm filters in place. On removing the filters (figure 12) the the effect is less visible but its width is reduced. The coincidence rate increases roughly fourfold because the filters transmit roughly 50% of incident 826.8$\pm$5 nm light. The beam radius at the crystal in this experiment was about 78 μm and $\theta = 14.3°$. Using the Gaussian narrowband approximation of eq. 61 we obtain a predicted halfwidth of 27 μm. A Gaussian fit to the experimental result has $1/e$ half width 19 μm (noting that the graphs shown are plotted as a function of $\delta x/2$ the distance moved by the translation stage). This is shorter than the simplified theory predicts most probably due to extra bandwidth introduced by the finite aperture size. A full theory of the effect including finite aperture effects is at present being prepared. As the coincidence rates were above 10 Hz for all results the experimental durations for each point measured were of order 1 minute and scans of the type shown took about 15 minutes to produce.

5 CONCLUSIONS

We have described several experiments that exploit the non-classical properties of spontaneous parametric downconversion. These experiments rely on the selection of two identical trains of photons by placing apertures in the cone of downconverted light in positions satisfying the phase matching conditions. A photo-electron event triggered shutter can be used in conjunction with a pair photon source so created to produce a gated photon train which is antibunched when there is a dead time between shutter open events. Within the limitations of the shutter this is also a source of single photon states. The limited shutter and detector efficiencies make this source only weakly sub-poissonian and the intensity is limited by the speed of the shutter.

An analogue feedback system, exploiting the higher efficiencies of solid state detectors has been used to create a sub-poissonian light source with a power of ~60pW and a Fano factor of 0.78. This is the lowest post detection Fano factor that we are aware of. In principle, with lower loss optics, optimised detectors and a crystal with higher non-linearity, noise reductions of a factor of 2 (3dB) at nanowatt powers should easily be achieved using this system. The bandwidth of the noise reduction is at present limited by the power of the source (10^8 photons/sec) however for sensitive absorption measurements bandwidths of a few kilohertz should be adequate in many situations. Pair photon sources based on optical parametric oscillators are described elsewhere in this book (Giacobino and Fabre, 1988). Such sources are more difficult to to stabilise but have the advantage of higher (mW) power and consequently higher bandwidth.

Practical noise reduction in absorption spectroscopy will only occur when some total absorbed power constraint applies. Given this power constraint pair photon sources can be used as long as effective detection efficiencies are high. For a simple subtractive estimator scheme the detector efficiecies must be greater than 50% for any gain to be acheived over a conventional measurement. However comparison with a conventional differential absorption scheme where the same dose limit holds for the monitor beam as for the probing beam shows noise reductions for all detector efficiences. Performing a differential intensity measurement of this type would lead to a noise reduction of about 50% using the present analogue detection system. Use of pair sources and coincidence techniques in the measurement of detector quantum efficiencies has also been demonstrated.

A theory of the two photon probability amplitude in parametric downconversion has been used to predict the length of a fourth order interference effect which occurs when identical photon 'pairs' are coherently combined at a beamsplitter. In such an experiment a sub-picosecond photon 'length' has been measured. This length is only limited by the physics of the parametric process itself. By suitably modifying the experiment it may be possible to directly probe this fundamental bandwidth. Control of photon overlap on the sub-picosecond timescale may be useful in the study of two photon absorption processes although a much higher photon flux may be necessary for any practical applications. The existence of this effect furthur emphasises the fact that the pair photons cannot be thought of as particles created at the crystal. The photon is better visualised as the smallest amount of energy extractable from the electromagnetic field falling on a detector. Tests of Bell's inequality based on the fourth order interference properties of photons have recently been suggested (Ou,1988).

Acknowledgements

The authors gratefully acknowledge contributions to this work from: E Jakeman, J S Satchell and J G Walker. We also acknowledge useful discussions with E R Pike, R G W Brown, K D Ridley and S Sarkar.

References

Brendel J, Schutrumpf S, Lange R, Martienssen W and Scully M O 1988, Europhys. Letts. 5 223

Brown R G W, Jakeman E, Pike E R, Rarity J G and Tapster P R 1986a, Europhys. Letts. 2 279

Brown R G W, Rarity J G and Ridley K D 1986b, Applied Optics 25 4122

Brown R G W, Jones R, Rarity J G and Ridley K D 1987, Applied Optics 26 2383

Burnham D C and Weinberg D L 1970, Phys. Rev. Lett. 25 84

Giacobino E and Fabre C 1988, this volume

Fearn H and Loudon R 1987, Opt. Commun. 64 485

Friberg S, Hong C K and Mandel L 1985, Phys. Rev. Lett. 54 2011

Gehrtz M, Bjorklund G C, Whittaker E A 1985, J. Opt. Soc. Am. B 2 1510

Hong C K and Mandel L 1987, Phys. Rev. Lett. 59 2044

Hong C K and Mandel L 1988, this volume

Heideman A, Horowicz R J, Reynaud S, Giaccobino E, Fabre C and Camy G 1987 Phys. Rev. Lett. 59 2555

Jakeman E and Rarity J G 1986, Opt. Commun. 56 219

Knoll G K 1979, Radiation detection and measurement, John Wiley, New York

Korde R and Geist J 1987, Applied Optics 26 5284

Machida S, Yamamoto Y and Itaya Y 1987, Phys. Rev. Lett., 58 1000

Mollow B R 1973, Phys. Rev. A 8 2684

Ou Z Y 1988, Phys. Rev. A 37 1607

Ou Z Y, Hong C K and Mandel L 1987, Opt. Commun. 63 118

Pike E R and Sarkar S 1988, this volume

Prasad S, Scully M O, Martienssen W 1987, Opt. Commun. 62 139

Rarity J G, Tapster P R and Jakeman E 1987, Opt. Commun., 62 201

Sergienko A V and Penin A N 1986, Sov. Tech. Phys. Lett. 12 328

Tapster P R, Rarity J G, and Satchell J S 1987, Europhys. Lett., 4,293

Tapster P R, Rarity J G, and Satchell J S 1988, Phys. Rev. A, in press.

Walker J G and Jakeman E 1984, Proc. SPIE 492 2

Walker J G and Jakeman E 1985, Opt. Acta 32 1303

Wong N C and Hall J L 1985, J. Opt. Soc. Am. B 2 1527

Yariv A 1976, Introduction to Quantum Electronics,Holt, Rinehart and Wilson, San Francisco, chapters 16,17

Yuen H P 1988, this volume.

MOVING MIRRORS AND NONCLASSICAL LIGHT

S SARKAR

1. INTRODUCTION

There is interest in very high precision measurements in connection with gravitational wave detection. Distances of the order of 10^{-21} m are involved, which are six orders of magnitude smaller than typical nuclear distances. As a consequence even intrinsic quantum fluctuations (ie noise) can be a severe limitation (Caves, 1980; Loudon, 1981) to the accuracy of the measurement. Much ingenuity has gone into trying to overcome some of these problems, one suggestion (Caves, 1981) being the use of squeezed states (Stoler, 1970; Yuen, 1976; Walls, 1986). The framework for the theoretical analysis of these quantum fluctuations in a (simplified) Michelson interferometer however is not entirely satisfactory. In Figure 1 the creation and annihilation operators for the two atoms $a_i^{\dagger}$ and a_i $(i = 1,2)$ are introduced just as in a standard cavity. As the mirrors are on springs they are free to move; they can move owing to impinging gravitational waves or the force due to radiation pressure. Since we no longer have a fixed cavity it is not clear as to what quanta in which modes are being created by the operators. For the gravitational wave detector the problem is compounded by the need for self-consistently determining the mirror position from the dynamics of the radiation field and the gravity waves, and of course the mirror position affects the radiation field. In this paper we want to isolate from these interrelated effects, the role of a given (ie predetermined) mirror motion on the radiation field within the cavity. (We assume the absence of gravitational waves; their effect may be thought of as being included in the predetermined motion of the mirror.) So far we have been tacitly assuming that the mirror is a mechanical one; however it may be possible to simulate the action of a mirror by using an optical nonlinearity to change the refractive index of a thin strip of dielectric. The effective movement of such a 'mirror' could then be very much faster than that of a mechanical one.

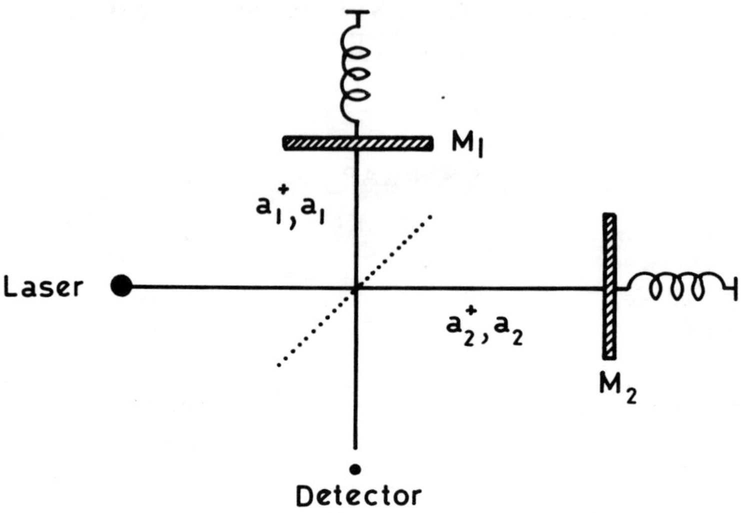

Figure 1. Michelson Interferometer

For simplicity we will consider not the Michelson interferometer geometry but a Fabry–Perot with only one mirror moving as shown in Figure 2.

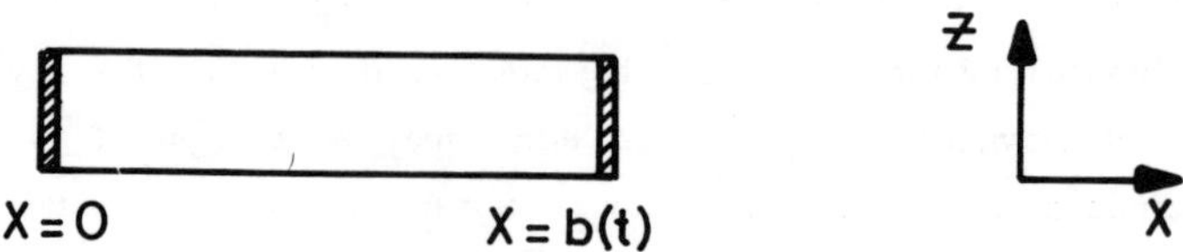

Figure 2. A Fabry–Perot Cavity

As an additional simplification we will regard the system as one dimensional (ie ignore transverse effects). Let us first consider the classical theory. If the electric field $\underline{E}(\underline{x},t)$ is linearly polarised in the z-direction then (in the Coulomb gauge) the electric and magnetic fields can be written as

$$\underline{E}(x,t) = -\frac{\partial A(x,t)}{\partial t}\,\underline{e}_z \tag{1}$$

$$\underline{B}(x,t) = \nabla \times (A(x,t)\,\underline{e}_z) \tag{2}$$

$\underline{e}_z$ is the unit vector in the z-direction and $A(x,t)\underline{e}_z$ is the vector potential. If we choose units so that the speed of light $c = 1$ then

$$\frac{\partial^2 A}{\partial t^2} = \frac{\partial^2 A}{\partial x^2} \tag{3}$$

In the instantaneous rest frame of the moving mirror (ie a frame moving with velocity $b(t)\underline{e}_x$ where $\underline{e}_x$ is the unit vector in the x-direction) the electric and magnetic fields are different. In particular the electric field $\underline{E}'$ is given by

$$\underline{E}' = \gamma(\underline{E}+\underline{v}\times\underline{B}) = -\gamma\left[\frac{\partial A}{\partial t} + \dot{b}(t)\frac{\partial A}{\partial x}\right]\underline{e}_z \tag{4}$$

where $\gamma = (1 - (\dot{b}(t))^2)^{-\frac{1}{2}}$. If the mirror is a perfect conductor then the tangential electric field in its rest frame vanishes and so

$$\frac{\partial A}{\partial t} + \dot{b}(t)\frac{\partial A}{\partial x} = 0 \tag{5}$$

This is equivalent to

$$\frac{d}{dt}A(b(t),t) = 0 \tag{6}$$

Without loss of generality we can then require that

$$A(b(t),t) = 0 \tag{7}$$

In particular if there is a vacuum then a unique solution within the cavity is

$$\underline{E} = \underline{B} = 0 \tag{8}$$

However in quantum theory we do not expect the situation to be so simple. The paradigm of the simple harmonic oscillator indicates that the Heisenberg uncertainty relations between canonically conjugate quantities would give some zero point energy (and motion). Even in a vacuum there are zero point fluctuations. There have to be charge and current fluctuations to maintain (7). From the solution of Maxwell's equations (the operator or c–number character of them (being irrelevant at this juncture) we know that accelerating charges and time varying currents can give rise to electromagnetic waves. Consequently in an empty cavity if the mirror is moved from rest some photons will be generated. One of the interesting aspects of these photons is that there is squeezing in some circumstances. If there are photons already present in the cavity, then the movement of the mirror will produce additional photons at different frequencies as well as quantum fluctuations. In order to see whether these expectations materialise we need to quantise the system.

2. QUANTISATION

In the traditional method of quantisation there is an important role for classical configuration space Q, which we take to be a differentiable manifold (Arnold, 1978). Two routes can be followed in the classical analysis both of course leading to the wave equation (3). For clarity let us consider first a finite dimensional system, ie one described by a finite number (N) of co–ordinates and their corresponding momenta. If we know the initial values of the co–ordinates and the tangent vector to the dynamical trajectory then we can predict the time evolution of the system. This is known as the Lagrangian formulation of the dynamics. Another approach, which is more closely linked to the quantum theory, is the canonical formulation. The state space is no longer the tangent bundle (Arnold, 1978) TQ but the bundle of 1 forms on Q or T^*Q, the cotangent bundle (Arnold, 1978). A 1–form at a point in Q is a linear functional on the vector space of tangent vectors at the same point. In the natural co–ordinate basis a general 1–form has coefficients $(p_1,p_2,\dots,p_N)$. These are the familiar momenta. Conventionally the co–ordinate basis of the 1–forms is written as $(dq^1,dq^2,\dots,dq^N)$, and so a general 1–form on TQ is

$$p_1dq^1 + p_2dq^2 + \dots + p_Ndq^N$$

This one form can also be regarded as a one form on the manifold T^*Q. For the dynamics of points on T^*Q we need a two form on T^*Q. This special form is known as a symplectic form ω^2 and is written as an exterior derivative

$$\omega^2 = -d\left[\sum_{i=1}^{N} p_i dq^i\right] = \sum_{i=1}^{N} dq^i \wedge dp_i \tag{9}$$

(This two form is a linear functional of any pair of tangent vectors at a point of T^*Q.) Any physical observable can be regarded as a smooth mapping from T^*Q to the $\mathbb{R}$, the set of all real numbers. Consequently an observable O can be written as a function O(q,p). Since it is customary to write the basis of tangent vectors of T^*Q as

$$\left[\frac{\partial}{\partial q^1}, \frac{\partial}{\partial q^2}, \ldots, \frac{\partial}{\partial q^N}, \frac{\partial}{\partial p_1}, \frac{\partial}{\partial p_2}, \ldots, \frac{\partial}{\partial p_N}\right]$$

we can associate the tangent vector field O'

$$O' = \sum_{i=1}^{N} \left[\frac{\partial O}{\partial p_i}\frac{\partial}{\partial q^i} - \frac{\partial O}{\partial q^i}\frac{\partial}{\partial p_i}\right] \tag{10}$$

to an observable O. Given any two observables O_1 and O_2 we can easily show that

$$\omega^2(O_1', O_2') = \sum_{i=1}^{N} \left[\frac{\partial O_1}{\partial q^i}\frac{\partial O_2}{\partial p_i} - \frac{\partial O_1}{\partial p_i}\frac{\partial O_2}{\partial q^i}\right] \tag{11}$$

This is just the familiar Poisson bracket; it is thus closely related to the existence of a certain symplectic form on the state space T^*Q. The usual process of going from the classical theory to the quantum theory is by replacing the Poisson bracket with $(i/\hbar)$ times the commutator.

The above discussion indicates to us that we should seek a symplectic structure to the space of real classical solutions (Moore, 1972) of the electromagnetic fields, or, equivalently, A(x,t) in the cavity. In order to have the field kinetic energy bounded we will require that A(t,x) is in the function space $H^1(I)$, and $\partial/\partial t$ A(t,x) is in the function space $L^2(I)$. Here I is the closed interval $\{x \mid 0 \leq x \leq b(t)\}$. L^2 is the set of functions which is square integrable on I and $H^1(I)$ is the set of functions whose x derivative is in $L^2(I)$. The elements of the cotangent bundle are

$$\left\{A(t,x) \quad , \quad \frac{\partial}{\partial t} A(t,x)\right\} \quad .$$

This is rather familiar since $\underline{A}$, and $\underline{E}$ are canonically conjugate in the standard approach to electromagnetism in the radiation gauge. Given two classical solutions $A^{(1)}(t,x)$ and $A^{(2)}(t,x)$ we can introduce the symplectic form

$$(A^{(1)}|A^{(2)}) = \int_0^{b(t)} dx \left\{A^{(2)}(t,x) \frac{\partial}{\partial t} A^{(1)}(t,x) - \left[\frac{\partial}{\partial t} A^{(2)}(t,x)\right] A^{(1)}(t,x)\right\} \tag{12}$$

Owing to (3) and (7) this form is independent of t. It is always possible to choose a set of real classical solutions $u_m(t,x)$ and $v_n(t,x)$ such that we have a form of 'orthogonality' relations

$$(u_m|v_n) = 0 = (v_m|v_n) \tag{13}$$

$$(u_m|v_n) = \delta_{mn} \tag{14}$$

Such a solution forms a basis for the cotangent bundle. (An explicit construction of the basis will be given presently.) Any solution A(t,x) can be written as

$$A(t,x) = \sum_n (\alpha_n v_n(t,x) - \beta_n u_n(t,x)) \tag{15}$$

where α_n and β_n are real numbers.

The u_n and v_n may also be regarded as observables since they induce maps into $\mathbb{R}$ from the cotangent bundle into $\mathbb{R}$, eg

$$u_n: \quad \left\{A, \frac{\partial A}{\partial t}\right\} \longrightarrow \left(\frac{\partial A}{\partial t} \,\middle|\, u_n\right)$$

$$v_n: \quad \left\{A, \frac{\partial A}{\partial t}\right\} \longrightarrow (A|v_n) \tag{16}$$

We can then associate operators p_n and q_n, in the quantum case, with these observables. In particular (Moore, 1972; Sarkar, 1988)

$$u_n \longrightarrow \hat{p}_n$$

$$v_n \longrightarrow \hat{q}_n \tag{17}$$

and

$$\left[\hat{p}_n, \hat{q}_m\right] = -i(u_n|v_m) = -i\delta_{nm} \tag{18}$$

(we have chosen units so that $\hbar = 1$.) If $R(\cdot)$ is a function which is twice differentiable and invertible and satisfies (Moore, 1972)

$$R(t-b(t)) = R(t+b(t)) - 2 \tag{19}$$

then it is easy to verify that

$$u_n(t,x) = \frac{1}{(2n\pi)^{\frac{1}{2}}} (\cos(n\pi R(t-x)) - \cos(n\pi R(t+x))) \tag{20a}$$

$$v_n(t,x) = \frac{1}{(2n\pi)^{\frac{1}{2}}} (\sin(n\pi R(t+x)) - \sin(n\pi R(t-x))) \tag{20b}$$

satisfy (13) and (14). In terms of the functions in (20) the field operator A(t,x) can be written as

$$\hat{A}(t,x) = \sum_n \left\{\hat{p}_n v_n(t,x) - \hat{q}_n u_n(t,x)\right\} \tag{21}$$

From (13) and (14) we deduce that

$$\hat{p}_m = (\hat{A}|u_m) \tag{22a}$$

$$\hat{q}_m = -(\hat{A}|v_m) \tag{22b}$$

If we consider now a join of two motions $b^{(1)}(t)$ and $b^{(2)}(t)$ at $t = t'$, then A can be written as

$$\hat{A}(t,x) = \sum_n \left\{\hat{p}_n^{(1)} v_n^{(1)}(t,x) - \hat{q}_n^{(1)} u_n^{(1)}(t,x)\right\} \quad \text{for} \quad t < t' \tag{23}$$

and

$$\hat{A}(t,x) = \sum_n \left\{\hat{p}_n^{(2)} v_n^{(2)}(t,x) - \hat{q}_n^{(2)} u_n^{(2)}(t,x)\right\} \quad \text{for} \quad t > t' \tag{24}$$

From the orthogonality relations at $t = t'$ we find

$$\hat{p}_n^{(2)} = \sum_m \left\{(v_m^{(1)}|u_n^{(2)})\,\hat{p}_m^{(1)} - (u_m^{(1)}|u_n^{(2)})\,\hat{q}_m^{(1)}\right\} \tag{25}$$

$$\hat{q}_n^{(2)} = \sum_m \left\{(v_m^{(1)}|v_n^{(2)})\,\hat{p}_m^{(1)} - (u_m^{(1)}|v_n^{(2)})\,\hat{q}_m^{(1)}\right\} \tag{26}$$

If we had another join to a different motion $b^{(3)}(t)$ at $t = t''$ ($> t'$) then we would have identical relations with $2 \to 3$ and $1 \to 2$. Clearly we can obtain in this way a linear relation between the (p^3,q^3) and (p^1,q^1). When such a relation is converted into one between creation and annihilation operators, we find that the destruction operator associated with motion 3 involves both destruction and creation operators for motion 1.

Consequently if we consider a sequence of motions: no motion, motion and no motion, then, even if there were no photons initially, in the final stationary cavity photons would have been generated. The physical reason for these photons as mentioned earlier are accelerating charges in the mirrors; this indicates that motion has to be relativistic in order to get large effects. Our analysis, hopefully, will indicate the magnitude of these effects together with qualitative features concerning the nature of the fluctuations of the photon states produced. The important quantities that need to be calculated are

$$(v_m^{(i)} | u_n^{(i+1)}) \quad , \quad (u_m^{(i)} | u_n^{(i+1)}) \quad ,$$

$$(v_m^{(i)} | v_n^{(i+1)}) \quad \text{and} \quad (u_m^{(i)} | v_n^{(i+1)})$$

for $i = 1,2$. Let us explicitly categorise the motions:

$$\begin{aligned} b(t) &= b^{(1)}(t) = L & t &\leqq 0 \\ &= b^{(2)}(t) & 0 < t &\leqq t_o \\ &= b^{(3)}(t) = b^{(2)}(t_o) = b_o & t_o &< t \end{aligned} \tag{27}$$

Moreover $b^{(2)}(o) = L$, $\dot{b}^{(2)}(o) = 0$ and $\dot{b}^{(3)}(t_o) = 0$. (Clearly we require $b(t) \geqq 0$ for all t.) For $b^{(1)}(t)$ and $b^{(3)}(t)$ it is easy to verify that

$$R^{(1)}(\xi) = \frac{\xi}{L} \tag{28}$$

and

$$R^{(2)}(\xi) = \frac{\xi}{b^{(2)}(t_o)} \tag{29}$$

Consequently we have

$$u_n^{(1)}(t,x) = \frac{1}{(2n\pi)^{\frac{1}{2}}}\left[\cos\frac{n\pi(t-x)}{L} - \cos\frac{n\pi(t+x)}{L}\right]$$

$$v_n^{(1)}(t,x) = \frac{1}{(2n\pi)^{\frac{1}{2}}}\left[\sin\frac{n\pi(t+x)}{L} - \sin\frac{n\pi(t-x)}{L}\right] \tag{30}$$

and

$$u_n^{(3)}(t,x) = \frac{1}{(2n\pi)^{\frac{1}{2}}}\left[\cos\left[\frac{n\pi(t-x)}{b_o}\right] - \cos\left[\frac{n\pi(t+x)}{b_o}\right]\right]$$

$$v_n^{(3)}(t,x) = \frac{1}{(2n\pi)^{\frac{1}{2}}}\left[\sin\left[\frac{n\pi(t+x)}{b_o}\right] - \sin\left[\frac{n\pi(t-x)}{b_o}\right]\right] \tag{31}$$

The important missing ingredient is an expression for $R^{(2)}(\xi)$. Moore (1972) has indicated a method to calculate this when $|\dot{b}^{(2)}(t)| << 1$. It does not require us to specify $b^{(2)}(t)$ in detail. $R^{(2)}$ is written in the form

$$R^{(2)}(t \pm x) = \int^t \frac{dt'}{b^{(2)}(t')} + g^{(\pm)}\left[\epsilon t\ ,\ \frac{x}{b^{(2)}(t)}\right] \tag{32}$$

ϵ being of order $|\dot{b}^{(2)}(t)|$. (For a stationary mirror problem $g^{(\pm)} = 0$.) The function $g^{(\pm)}(\xi,t)$ satisfies the partial differential equation

$$(1 \pm \xi\dot{b}^{(2)}(t))\frac{\partial}{\partial\xi}g^{\pm}(t,\xi) \mp \epsilon b^{(2)}(t)\frac{\partial g^{\pm}}{\partial t}(t,\xi) \mp 1 = 0 \tag{33}$$

since

$$\frac{\partial}{\partial t}R^{\pm}(t,x) = \pm\frac{\partial R^{\pm}}{\partial x}(t,x) \tag{34}$$

and

$$R^{\pm}(t,x) = R(t \pm x) \tag{35}$$

We can expand $g^{\pm}(t,\xi)$ as

$$g^{\pm}(t,\xi) = \sum_{n=0}^{\infty} g_n^{\pm}(t,\xi)\ \epsilon^n \tag{36}$$

and

$$g_n^{\pm}(t,\xi) = \sum_{j=0}^{n+1} g_{n+1,j}^{\pm}(t)\ (\mp\ \xi)^j \tag{37}$$

On equating powers of ϵ in (33) it is possible to determine $g^{\pm}(t,\xi)$ and hence $R^{(2)\pm}(t,\xi)$ for a general motion $b^{(2)}(t)$. We can show that (Sarkar, 1988)

$$R^{(2)\pm}(t,x) = a(t,x) + c^{\pm}(t,x) + O(\epsilon^3) \tag{38}$$

where

$$a(t,x) = \int^{t} \frac{dt'}{b^{(2)}(t')} + \frac{1}{6}\int^{t} dt' \left[-2\,\frac{[\dot{b}^{(2)}(t')]^2}{b^{(2)}(t')} + \ddot{b}^{(2)}(t')\right]$$

$$- \tfrac{1}{2}\,\frac{x^2}{[b^{(2)}(t)]^2}\,\dot{b}^{(2)}(t) \tag{39}$$

and

$$c^{\pm}(t,x) = \pm\left[\frac{x}{b^{(2)}(t)} + \frac{1}{6}\left[\frac{x}{b^{(2)}(t)} - \frac{x^3}{[b^{(2)}(t)]^3}\right]\left[-2\,[\dot{b}^{(2)}(t)]^2 + \right.\right.$$

$$\left.\left. b^{(2)}(t)\,\ddot{b}^{(2)}(t)\right]\right] \tag{40}$$

It should be noted that in (39) there is no lower limit on the integral. A constant α can be added to $R^{(2)}$ without affecting the physical content of the theory. Indeed if $R^{(2)} \rightarrow R^{(2)} + \alpha$ then

$$u_n^{(2)} \longrightarrow u_n^{(2)'} = \cos(n\pi\alpha)\, u_n^{(2)} + \sin(n\pi\alpha)\, v_n^{(2)} \tag{41a}$$

$$v_n^{(2)} \longrightarrow v_n^{(2)'} = \cos(n\pi\alpha)\, v_n^{(2)} - \sin(n\pi\alpha)\, u_n^{(2)} \tag{41b}$$

This is an orthogonal transformation and the $\hat{p}'_n$, $\hat{q}'_n$ have the same commutation relations as $\hat{p}_n$, $\hat{q}_n$.

3. GENERATION OF PHOTONS

In this section various statistical quantities characterising photons produced by moving mirrors will be calculated.

Using the procedure outlined in section 2 we find

$$\hat{p}_n^{(2)} = -\frac{1}{2L}\sum_m \left[\frac{m}{n}\right]^{\frac{1}{2}} \left[I^+_{mn}(L,\beta) + I^-_{mn}(L,\beta)\right] \hat{p}_m^{(1)} + O(\epsilon^3) \tag{42}$$

and

$$\hat{q}_n^{(2)} = -\frac{1}{2L}\sum_m \left[\frac{m}{n}\right]^{\frac{1}{2}} \left[I^-_{mn}(L,\beta) - I^+_{mn}(L,\beta)\right] \hat{q}_m^{(1)} + O(\epsilon^3) \tag{43}$$

where

$$I^{\pm}_{kn}(L,\gamma) = \int_{-L}^{L} dx \cos\pi \left[(k\pm n)\frac{x}{L} \pm n\gamma\left[\frac{x}{L} - \frac{x^3}{L^3}\right]\right] \tag{44}$$

and

$$\beta = \frac{1}{6}\, \ddot{b}^{(2)}(o)\, L$$

Similarly

$$\hat{p}_n^{(3)} = \frac{1}{2b_o} \sum_{m'} \left\{ \cos\left[\frac{\pi n t_o}{b_o}\right] \left[\frac{n}{m'}\right]^{\frac{1}{2}} \left[I^+_{nm'}(b_o, d_o) - I^-_{nm'}(b_o, d_o)\right] \hat{p}_{m'}^{(2)} \right.$$

$$\left. - \sin\left[\frac{\pi n t_o}{b_o}\right] \left[\frac{n}{m'}\right]^{\frac{1}{2}} \left[I^+_{nm'}(b_o, d_o) + I^-_{nm'}(b_o, d_o)\right] \hat{q}_{m'}^{(2)} \right\}$$

$$+ O(\epsilon^3) \tag{45}$$

and

$$\hat{q}_n^{(3)} = \sum_{m'} \left\{ \frac{1}{2b_o} \left[\frac{n}{m'}\right]^{\frac{1}{2}} \sin\left[\frac{\pi n t_o}{b_o}\right] \left[I^-_{nm'}(b_o, d_o) - I^+_{nm'}(b_o, d_o)\right] \hat{p}_{m'}^{(2)} \right.$$

$$\left. - \frac{1}{2b_o} \left[\frac{n}{m'}\right]^{\frac{1}{2}} \cos\left[\frac{\pi n t_o}{b_o}\right] \left[I^+_{nm'}(b_o, d_o) + I^-_{nm'}(b_o, d_o)\right] \hat{q}_{m'}^{(2)} \right\}$$

$$+ O(\epsilon^3) \tag{46}$$

d_o denotes $b_o\ \ddot{b}^{(2)}(t_o)/6$. It is easy to check that to $O(\epsilon^3)$ the above expressions preserve commutation relations. Using (45), (42) and (43) $\hat{p}_n^{(3)}$ and $\hat{q}_n^{(3)}$ can be expressed in terms of $\hat{p}_n^{(1)}$ and $\hat{q}_n^{(1)}$. It is more convenient at this stage to work in terms of creation and annihilation operators which are given in the usual way by

$$a_n^+ = \frac{1}{\sqrt{2}} (\hat{q}_n - i\hat{p}_n)$$

$$a_n = \frac{1}{\sqrt{2}} (\hat{q}_n + i\hat{p}_n) \tag{47}$$

We can show straightforwardly that

$$a_n^{(3)} = \tfrac{1}{2} \sum_m \Big[a_m^{(1)} \Big[u_m^{(1)}(n) + t_m^{(1)}(n) + i \Big[v_m^{(1)}(n) - s_m^{(1)}(n) \Big]\Big]$$

$$+ a_m^{(1)+} \Big[t_m^{(1)}(n) - u_m^{(1)}(n) + i \Big[v_m^{(1)}(n) + s_m^{(1)}(n) \Big]\Big]\Big] \qquad (48)$$

where

$$s_m^{(1)}(n) = -\frac{1}{4b_o L} \sum_{m'} \left[\frac{n}{m'}\right]^{\frac{1}{2}} \left[\frac{m}{m'}\right]^{\frac{1}{2}} \sin\left[\frac{\pi n t_o}{b_o}\right] \cdot \left[I^-_{nm'}(b_o, d_o) - I^+_{nm'}(b_o, d_o) \right]$$

$$\cdot \left[I^+_{mm'}(L, \beta) + I^-_{mm'}(L, \beta) \right] \qquad (49)$$

$$t_m^{(1)}(n) = \frac{1}{4b_o L} \sum_{m'} \left[\frac{n}{m'}\right]^{\frac{1}{2}} \left[\frac{m}{m'}\right]^{\frac{1}{2}} \cos\left[\frac{\pi n t_o}{b_o}\right] \left[I^+_{nm'}(b_o, d_o) + I^-_{nm'}(b_o, d_o) \right]$$

$$\cdot \left[I^-_{mm'}(L, \beta) - I^+_{mm'}(L, \beta) \right] \qquad (50)$$

and

$$u_m^{(1)}(n) = -\cot\left[\frac{n\pi t_o}{b_o}\right] s_m^{(1)}(n) \qquad (51a)$$

$$v_m^{(1)}(n) = \tan\left[\frac{n\pi t_o}{b_o}\right] t_m^{(1)}(n) \qquad (51b)$$

We have now all the information to calculate interesting quantities. If the initial state in the cavity is a vacuum then the mean number of photons created in mode n is

$$<o| a_n^{(3)+} a_n^{(3)} |o>$$

$$= \frac{1}{4} \sum_m <1| \left[t_m^{(1)}(n) - u_m^{(1)}(n) - i \left[v_m^{(1)}(n) + s_m^{(1)}(n) \right] \right]$$

$$\cdot \left[t_m^{(1)}(n) - u_m^{(1)}(n) + i \left[v_m^{(1)}(n) + s_m^{(1)}(n) \right] \right] |1>$$

$$= \frac{1}{4} \sum_m \left[\left[t_m^{(1)}(n) - u_m^{(1)}(n) \right]^2 + \left[v_m^{(1)}(n) + s_m^{(1)}(n) \right]^2 \right] \quad (52)$$

No photons are produced if

$$t_m^{(1)}(n) = u_m^{(1)}(n) \quad \text{and} \quad v_m^{(1)}(n) = -s_m^{(1)}(n) \quad .$$

(As a check on the consistency of our calculations we note that if the mirror is always stationary (ie $b^{(2)}(t)$ is constant) then

$$u_m^{(1)}(n) = \cos \left[\frac{n\pi t_o}{b_o} \right] \delta_{nm} = t_m^{(1)}(n)$$

$$(53)$$

$$v_m^{(1)}(n) = \sin \left[\frac{n\pi t_o}{b_o} \right] \delta_{nm} = -s_m^{(1)}(n)$$

and so no photons are indeed produced.) In general, on using

$$I_{kn}^{(\pm)}(L,\gamma) = 2L \, \delta_{k,\mp n} \pm (-1)^{k\pm n} (1 - \delta_{k,\mp n}) \frac{12n\gamma L}{\pi^2 (k\pm n)^3} \quad (54)$$

we find that

$$<o| a_n^{(3)+} a_n^{(3)} |o> = \frac{36n^2}{\pi^4} \sum_{m=1}^{\infty} \left[\left[\frac{n}{m} \right]^{\frac{1}{2}} d_o - \left[\frac{m}{n} \right]^{\frac{1}{2}} \beta \right]^2 \left[\frac{1}{(n+m)^3} - \frac{(1-\delta_{nm})}{(n-m)^3} \right]^2 \quad (55)$$

This is not compatible with a thermal distribution for which we would have

$$\langle o| \; a_n^{(3)+} a_n^{(3)} \; |o\rangle \sim \frac{1}{\exp\left[\frac{n}{T'}\right] - 1} \tag{56}$$

and T' would be proportional to the temperature. In fact (55) implies a power law fall-off with n.

Now we will calculate the variance of the distribution of photons in order to determine whether it is Poissonian, super-Poissonian or sub-Poissonian. We find that

$$\langle o| \; a_n^{(3)+} a_n^{(3)+} a_n^{(3)} a_n^{(3)} \; |o\rangle - \left[\langle o| \; a_n^{(3)+} a_n^{(3)} \; |o\rangle\right]^2$$

$$= \frac{1}{16\sin^4\left[\frac{n\pi t_o}{b_o}\right]} \left[\sum_{m,m'} \left[s_m^{(1)}(n) + \tan\left[\frac{\pi n t_o}{b_o}\right] t_m^{(1)}(n)\right]^2\right.$$

$$\cdot \left[s_{m'}^{(1)}(n) + \tan\left[\frac{\pi n t_o}{b_o}\right] t_{m'}^{(1)}(n)\right]^2$$

$$\left. + \sum_m \left[\tan^2\left[\frac{\pi n t_o}{b_o}\right] \left[t_m^{(1)}(n)\right]^2 - \left[s_m^{(1)}(n)\right]^2\right]^2\right] \tag{57}$$

Since the right hand side of this equation is a sum of positive terms we have super-Poissonian statistics.

In order to discuss a phase sensitive quantity such as squeezing we consider the operator

$$X_n(\theta) = e^{-i\theta} a_n^{(3)} + e^{i\theta} a_n^{(3)+} \tag{58}$$

Clearly

$$\langle o| \; X_n(\theta) \; |o\rangle = 0 \tag{59}$$

We can calculate the variance and find

$$<o\left|(X_n(\theta))^2\right|o> - \left[<o\left|X_n(\theta)\right|o>\right]^2$$

$$= \tfrac{1}{2}\cos\left[2\left[\frac{n\pi t_o}{b_o} - \theta\right]\right]\sum_m\left[\frac{1}{\cos^2\frac{n\pi t_o}{b_o}}\left[t_m^{(1)}(n)\right]^2 - \frac{1}{\sin^2\frac{n\pi t_o}{b_o}}\left[s_m^{(1)}(n)\right]^2\right]$$

$$+ \tfrac{1}{2}\sum_m\left[\left[t_m^{(1)}(n) - u_m^{(1)}(n)\right]^2 + \left[v_m^{(1)}(n) + s_m^{(1)}(n)\right]^2\right] \qquad (60)$$

$$+ 1$$

If we write

$$\frac{t_m^{(1)}(n)}{\cos\left[\frac{n\pi t_o}{b_o}\right]} = \delta_{nm} + a_{nm} \qquad (61)$$

and

$$\frac{s_m^{(1)}(n)}{\sin\left[\frac{n\pi t_o}{b_o}\right]} = -\delta_{nm} + b_{nm} \qquad (62)$$

(with $a_{nm} - O(\epsilon^2)$ and $b_{nm} = O(\epsilon^2)$) then

$$<o\left|(X_n(\theta))^2\right|o> \simeq 1 + (a_{nn}+b_{nn})\cos 2\left[\frac{n\pi t_o}{b_o} - \theta\right] \qquad (63)$$

Similarly

$$<o\left|(X_n(\theta+\pi/2))^2\right|o> \simeq 1 - (a_{nn}+b_{nn}) \cos 2\left[\frac{n\pi t_o}{b_o} - \theta\right] \tag{64}$$

To the accuracy that we are working

$$(<o\left|(X_n(\theta))^2\right|o> - \left[<o\left|X_n(\theta)\right|o>\right]^2)$$

$$(<o\left|\left[X_n\left[\theta + \frac{\pi}{2}\right]\right]^2\right|o> - \left[<o\left|X_n\left[\theta + \frac{\pi}{2}\right]\right|o>\right]^2)$$

$$= 1 \tag{65}$$

and so we have a minimum uncertainty state. From (63) we see that it is possible to have squeezing if

$$(a_{nn}+b_{nn}) \cos 2\left[\frac{n\pi t_o}{b_o} - \theta\right] > 0 \tag{66}$$

From (49), (50) and (54) we can deduce that

$$a_{nm} = \frac{-6n}{\pi^2(n-m)^3} (-1)^{m-n} (1-\delta_{nm}) \left[\left[\frac{n}{m}\right]^{\frac{1}{2}} d_o - \left[\frac{m}{n}\right]^{\frac{1}{2}} \beta\right]$$

$$+ \frac{(-1)^{n+m}\, 6n}{\pi^2(n+m)^3} \left[\left[\frac{n}{m}\right]^{\frac{1}{2}} d_o - \left[\frac{m}{n}\right]^{\frac{1}{2}} \beta\right] \tag{67}$$

and

$$b_{nm} = (-1)^{m-n} \frac{(1-\delta_{nm})}{\pi^2} \frac{6n}{(m-n)^3} \left[\left[\frac{m}{n}\right]^{\frac{1}{2}} \beta - \left[\frac{n}{m}\right]^{\frac{1}{2}} d_o\right]$$

$$- \frac{(-1)^{m+n}\, 6n}{\pi^2(m+n)^3} \left[\left[\frac{m}{n}\right]^{\frac{1}{2}} \beta - \left[\frac{n}{m}\right]^{\frac{1}{2}} d_o\right] \tag{68}$$

Consequently (66) would imply

$$\frac{1}{4\pi^2 n^2} \cos 2\left[\frac{n\pi t_o}{b_o} - \theta\right]\left[b_o \ddot{b}(t_o) - L\ddot{b}(o)\right] > 0 \qquad (69)$$

Provided $(b_o\ddot{b}(t_o) - L\ddot{b}(o)) > 0$ we can arrange this and so have squeezing for some values of θ.

A particularly simple form for $b^{(2)}(t)$ is simple harmonic motion

$$b^{(2)}(t) = c_o + c_1 \cos\omega t \qquad (70)$$

with $c_o + c_1 = L$ and c_o, $c_1 > 0$. Now for this motion

$$b_o \ddot{b}(t_o) - L\ddot{b}(o) = \omega^2 c_1 (1 - \cos\omega t_o)(L + c_1 \cos\omega t_o) \qquad (71)$$

For $\omega t_o = \pi$ this is non-zero, but for a full period (ie $\omega t_o = 2\pi$) this is zero. Consequently, in this case, we can only have a squeezed photon state if the motion consists of half a cycle of sinusoidal motion. For complete cycles there is no squeezing. The squeezing parameter (Walls and Reid, 1986) is easily calculated to be

$$\frac{\omega^2 c_1 (L-c_1)}{4\pi^2 n^2} \cos^2\left[\frac{n\pi^2}{\omega(L-2c_1)} - \theta\right] \qquad (72)$$

Although taking the vacuum state to be the initial state is somewhat intriguing, the case when there are some <u>photons in the initial state</u> may be easier to realise. For definiteness we consider the state initially to be

$$|n'_o\rangle = \left[a_{n_o}^{(1)+}\right]^{n'_o} |o\rangle \qquad (73)$$

n_o is the mode number while n'_o is the number in the mode. The mean number of photons in the state generated is calculated as before and we find

$$<n_o'| a_n^{(3)+} a_n^{(3)} |n_o'> \simeq n_o' \left[\delta_{nn_o} + \frac{36n^2}{\pi^4 (n-n_o)^6} (1-\delta_{nno}) \left[\left[\frac{n}{n_o}\right]^{\frac{1}{2}} d_o - \left[\frac{n_o}{n}\right]^{\frac{1}{2}} \beta\right]^2\right] \tag{74}$$

So if n_o' is large enough the number of photons created in modes with mode number (n ($\neq n_o$) is proportionately large. In some sense the presence initially of photons is stimulating the generation of new ones (but not at the same mode number). Since

$$<n_o'| X_n(\theta) |n_o'> = 0 \tag{75}$$

and

$$<n_o'| (X_n(\theta))^2 |n_o'> \simeq \cos\left[2\left[\frac{n\pi t_o}{b_o} - \theta\right]\right] (a_{nn}+b_{nn}) (1 + 2\delta_{nn_o} n_o')$$

$$+ 1 + 2n_o' (1 + a_{nn_o} - b_{nn_o}) \delta_{nn_o} \tag{76}$$

the product of the variances in different quadratures is

$$<n_o'| (X_n(\theta))^2 |n_o'><n_o'| \left[X_n\left[\theta + \frac{\pi}{2}\right]\right]^2 |n_o'> \simeq \left[1 + 2n_o' \delta_{n_o n}\right]^2 \tag{77}$$

For mode numbers other than n_o the minimum uncertainty condition is satisfied. From (76) we see that the discussion of squeezing for photons in these mode numbers goes through as before.

For all mechanical mirrors the treatment we have given should be adequate. If non-mechanical mirrors can be designed to have relativistic motion then corrections will be necessary although our hope is that qualitative features will remain unchanged. To check this a non-perturbative calculation of $R^{(2)}$ is needed. For the case of the motion in (70) we can take

$$R^{(2)}(t) = \sum_n r_n e^{in\omega t} \tag{78}$$

Substituting in (19) we obtain

$$e^{iu\omega t} \sum_n r_n (-i)^{u-n} \left[e^{in\omega c_o} J_{n-u}(n\omega c_1) - e^{-in\omega c_o} J_{n-u}(-n\omega c_1)\right] = 2\delta_{uo} \tag{79}$$

There is one such relation for each integer u. Preliminary investigations suggest that such an expansion may be convergent. However the calculation of the mean number of photons, etc, will still be difficult.

We conclude by giving some estimates of the orders of magnitude of the effects that we are predicting. For the case of a full cycle of periodic motion

$$\langle o | a_1^{(3)+} a_1^{(3)} | o \rangle = \frac{1}{\pi^4} (\ddot{b}^{(2)}(o)L)^2 \sum_{m=1}^{\infty} \left[\frac{1}{m^{\frac{1}{2}}} - m^{\frac{1}{2}}\right]^2 \left[\frac{1}{(m+1)^3} - \frac{(1-\delta_{1m})}{(1-m)^3}\right]^2$$

$$\sim \frac{1}{\pi^4} r^2 \epsilon^4 \tag{80}$$

where $r = L/c_1$ and $\epsilon = c_1\omega/c$. (Here we have changed units and reintroduced c the speed of light which we took to be 1 before.) Similarly

$$\langle n'_o | a^{(3)+}_{n_o+1} a^{(3)}_{n_o+1} | n'_o \rangle \sim \frac{n'_o}{\pi^4} r^2 \epsilon^4 \left[1 + \frac{1}{n_o}\right] \tag{81}$$

A similar result holds for the mode (n_o-1). In order to be able to observe these photons the thermal background has to be low and so we would need

$$\frac{1}{\exp\left[\frac{\hbar c \pi}{kT\ (c_o+c_1)}\right] - 1} \sim \frac{1}{\pi^4} \epsilon^4 r^2 \tag{82}$$

or for the case of (81)

$$\frac{1}{\exp\left[\frac{\hbar c \pi}{kT\ (c_o+c_1)}\right] - 1} \sim \frac{n'_o}{\pi^4} \epsilon^4 r^2 \tag{83}$$

For the case $c_o = 0.1$ m, $c_1 = 0.05$ m and $\omega = 6 \times 10^8$ sec^{-1} we have that for an initial vacuum state the rate of photon generation per second is 6700. If we choose for an initial $|n'_o\rangle$ state, $n'_o \sim 9 \times 10^4$, then T is large and the average number of photons produced per second is 6×10^8, a substantial increase. Such an effect may be experimentally detectable. For the case when the initial state is a vacuum the squeezing parameter is $(6 \times 10^{-4})/n^2$. For an initial state $|n'_o\rangle$ this result continues to hold for mode numbers n different from n_o.

4. CONCLUSIONS

We have shown that in general motion of a mirror (from rest to eventual rest) will create photons, even if the initial state is a vacuum. In particular the photons generated from the vacuum are super–Poissonian, non–thermal, and in certain circumstances squeezed. Unless the mirror motion is effectively relativisitic the effects are small. Without going to extremely fast motions the amount of photon generation is much enhanced if the initial state is a pure number state with a large population. Experiments checking these predictions would be important in deepening our understanding of quantum field theory in non–trivial geometries.

ACKNOWLEDGEMENTS

I am grateful to J.G. Rarity, J.S. Satchell and P.R. Tapster for discussions.

REFERENCES

Arnold, V.I. 1978, Mathematical Methods of Classical Mechanics, Springer–Verlag, New York.
Caves, C.M. 1980, Phys. Rev. Lett. 45 75.
Caves, C.M. 1981, Phys. Rev. D23 1693.
Loudon, R. 1981, Phys. Rev. Lett. 47 815–818.
Moore, G.T. 1970, J. Math. Phys. 11 2679–1691.
Sarkar, S. 1988, J. Phys. A21 971–980.
Stoler, D. 1970, Phys. Rev. D1 3217.
Yuen, H.P. 1976, Phys. Rev. A13 2226.
Walls, D.F. and Reid, M.D. 1986, Frontiers in Quantum Optics, Ed. E.R. Pike and S. Sarkar, Hilger, Bristol, 72–105.

PROPAGATION OF NONCLASSICAL LIGHT

I ABRAM

1. INTRODUCTION

Classical optics describes quite successfully the propagation of laser light, both in free-space and inside a transparent medium. The reason is that in a coherent state of radiation, the electric and magnetic fields may be written in terms of their expectation values, and thus their propagation may be treated classically through the macroscopic Maxwell equations. The same holds also for thermal light which can always be described as a statistical mixture of coherent states of the electromagnetic field. The classical Maxwell equations permit the calculation of both the spatial progression and the temporal evolution of a propagating electromagnetic field, and treat the interaction of the field with a medium phenomenologically through the induced polarization.

On the other hand, neither classical optics nor the conventional formulation of quantum optics can address adequately the propagation of nonclassical light. The failure of classical optics in this problem is not surprising: The expectation value of the electric field in a nonclassical radiation state, such as a single-photon state or a squeezed state, does not characterize the photon statistics. Thus, its use in the Maxwell equations cannot give any information on the evolution of the photon statistics as light propagates through linear or nonlinear media. The inadequacy of conventional quantum optics is less obvious, and lies in the fact that this theory is based on the Hamiltonian of the radiation modes. It thus addresses only the time-evolution of the field propagating in (mostly) empty space and interacting with individual material points representing emitters, absorbers or polarizable entities that cause (nonlinear) interactions among the radiation modes. A theory based on a modal Hamiltonian, however, cannot address directly the problem of the spatial progression of the field operators propagating through macroscopic, translationally-invariant, polarizable media, since it cannot produce spatial differential equations which could account for the change in mode structure and in the excitation of each mode, as different media are introduced in the path of a light beam. For this reason, one of the most basic problems of classical propagative optics, the spatial progression of light through a transparent refractive medium, and the reflection and refraction that occurs as light crosses a vacuum/dielectric interface, cannot be treated at all within the conventional formalism of quantum optics: There is no interaction term that, when added to the free-field Hamiltonian, can describe the modification of the spatial progression of the field due to refraction.

Nonclassical states of light are usually described by second-quantization operators acting onto the vacuum state $|0\rangle$ of the radiation field, and the corresponding expectation values of the electric and magnetic fields are often zero. Clearly, the description of propagation of nonclassical light requires a direct quantum mechanical treatment of the spatial progression of the field operators. Such a treatment has been developed within the formalism of Propagative Quantum Optics (Abram, 1987; Abram, 1988), which can address the problem of the quantum mechanical description of the spatial progression of light independently of its temporal evolution.

In the following Sections, we outline the main features of propagative quantum optics (Section 2), and examine more particularly the treatment of propagation through a linear (refractive) medium (Section 3), and across a refractive interface (Section 4). In order to facilitate the discussion of the physical ideas, we use a plane-wave propagation geometry which greatly simplifies the mathematical treatment.

We thus consider electromagnetic plane waves propagating along the $\pm$z-axis in a refractive medium, with the electric field E and the material polarization P both along the x-axis, and the magnetic field H and the magnetic induction B along the $\pm$y-axis. We consider nonmagnetic materials, so that B = H. In this geometry we can treat all quantities as scalars, their direction being always implicit. We use Gaussian units. Quantum mechanical operators are distinguished from the corresponding classical quantities by the caret ($\hat{}$) and we take $\hbar = 1$.

2. BASIC CONSIDERATIONS OF PROPAGATIVE QUANTUM OPTICS

The description of light propagation requires consideration of the variation of the electromagnetic field along (at least) two dimensions: time plus (at least) one spatial coordinate. Conventional quantum optics (like most quantum mechanical theories) gives a central role to the Hamiltonian of the electromagnetic field, and thus within this theory only temporal differential equations can be set up (i.e. the equations of motion of the field that can be obtained from the Heisenberg equation), through which the time-evolution of the normal modes of the field can be calculated. Clearly, a complete quantum mechanical description of light propagation requires an extension of quantum optics to include explicit calculations of the spatial progression of the field, through direct spatial differential equations. This can be achieved quite easily by use of the momentum operator of the electromagnetic field $\hat{G}$, since by definition

$$\hat{G}\,\Psi = -i\frac{\partial}{\partial z}\Psi \tag{2.1}$$

or equivalently

$$\frac{\partial}{\partial z}\hat{Q} = -i\,[\hat{G},\hat{Q}] \tag{2.2}$$

where Ψ is any wavefunction and $\hat{Q}$ any operator.

In calculating the Hamiltonian and momentum operators, the field is treated quantum mechanically through second-quantization operators. On the other hand, the medium is introduced through its induced polarization, or equivalently through its optical susceptibility $\chi = P/E$. This quasi-phenomenological treatment of the material medium permits separation of the problem of propagation of the field from the microscopic description of the field-matter interaction. The optical susceptibility can be calculated beforehand (for example through the diagrammatic techniques developed in nonlinear optics by Yee and Gustafson (1978)) and then used simply as a characteristic constant of the medium in the treatment of propagation. Thus, within this approach, light propagation is treated in a manner completely analogous to the macroscopic Maxwell equations, on which is based classical propagative optics. This similarity suggests that the results of classical optics may serve as a guide in choosing the appropriate quantum mechanical energy and momentum operators that describe light propagation properly.

From classical optics, we know that when a harmonic light wave enters a transparent refractive medium, its time-evolution does not change: it remains time-harmonic at the same frequency. Equivalently, the energy of the electromagnetic field propagating through different media, remains all in the field, and is independent of the medium. This is a consequence of energy conservation, since a transparent material does not extract any energy from the electromagnetic field propagating through it. On the other hand, the electromagnetic energy-density, given by

$$u = \frac{1}{8\pi}\left(E^2 + H^2 + 4\pi PE\right) \tag{2.3}$$

is larger inside the refractive medium (than in free space), as it includes the effects of the induced polarization. For electric and magnetic fields consistent with the Maxwell equations, the energy-density is proportional to the refractive index of the medium ($n = \sqrt{1 + 4\pi\chi}$), and is thus related to the free-field energy by

$$u_R = n\, u_o \qquad (2.4)$$

where the subscripts R and o refer to the density in the refractive medium and in empty space respectively. This means that the quantum mechanical description of light propagation must be based on a Hamiltonian that is independent of the medium of propagation, but on an energy-density operator whose eigenvalues are proportional to the refractive index. Since the Hamiltonian and energy-density operators are related to each other by integration over the volume V of the cavity of quantization,

$$\hat{\mathcal{H}} = V\, \hat{u} \qquad (2.5)$$

this implies that energy-conservation upon propagation can be introduced in quantum optics through the appropriate definition of the cavity of quantization, as will be seen later.

At the same time, introduction of a refractive medium in the path of a light beam, modifies its spatial progression: the wavelength of a harmonic wave is reduced by the refractive index of the medium. Equivalently, the momentum (wavevector) of a harmonic wave depends on the medium of propagation and is proportional to the refractive index. This in turn means that the appropriate momentum operator that describes the spatial progression of the field quantum mechanically depends on the medium, and its eigenvalues (the wavevectors) are proportional to the refractive index. In classical electrodynamics there are several ways of defining the field momentum inside a material medium, because of the different ways that the "total" momentum may be partitioned between "electromagnetic" and "mechanical" momentum. The definition that conforms to the above requirements is the so-called Minkowski momentum, and corresponds essentially to a pseudo-momentum that describes the spatial invariance of the field inside a stationary medium (Peierls, 1985). The Minkowski momentum density is a vector directed along the z-axis, whose value is given by

$$g = \frac{1}{4\pi c} D\ B \tag{2.6}$$

where

$$D = E + 4\pi P = (1 + 4\pi\chi)\ E \tag{2.7}$$

is the electric displacement.

We turn now to the cavity of quantization. This cavity is usually taken to be a finite box, whose length is an integral number of wavelengths of the radiation waves of interest, but large enough so that its mode spacing is beyond experimental resolution. At opposite sides of the box, the field obeys periodic boundary conditions and thus the cavity constitutes a "representative unit" whose repetition in three dimensions can cover all of (infinite) space. In order to study the field in infinite space, it is thus sufficient to examine one "representative unit". When a transparent refractive mendium is introduced in (infinite) space, the dimensions of the periodic-boundary cavity change since all wavelengths become shorter by the refractive index, and the planes on which periodic boundary conditions are obeyed get closer by the same factor. Thus, the volume of the periodic-boundary cavity filled with a refractive medium, is

$$V_R = \frac{V_o}{n} \tag{2.8}$$

This equation, combined with eq. (2.4) indicates that if we require that periodic-boundary conditions be obeyed by the cavity of quantization, independently of its material contents, then its energy contents are also independent of the medium. That is,

$$V_R u_R = V_o u_o \tag{2.9}$$

and thus the Hamiltonian (and the time-evolution) of the electromagnetic field quantized in such a cavity is the same, whether the cavity is empty or contains (one or more) refractive media. The energy conservation feature of propagative optics is thus equivalent to the preservation of the periodic-boundary conditions used for the cavity of quantization in quantum optics.

Similarly, the Minkowski momentum contained in the periodic-boundary cavity is,

$$G = V_R g_R = \frac{V_o}{n} (1 + 4\pi\chi)\ E\ B = n\ V_o g_o \tag{2.10}$$

Thus, according to this definition, the momentum contents of the cavity are proportional to the refractive index, and thus they depend on the material contents of the cavity in the same way as the wavevector. This implies that the quantum mechanical operator that corresponds to the Minkowski momentum should describe successfully the spatial progression of the field in any medium.

3. PROPAGATION IN A REFRACTIVE MEDIUM

The discussion of Section 2 may easily be applied to the quantized electromagnetic field. The electric and magnetic field operators quantized in free space may be written as (Abram, 1987)

$$\hat{E}(z,t) = \sum_j \hat{e}_j = \sum_j -i\left(\frac{2\pi\omega_j}{V_o}\right)^{1/2} (\hat{b}_j^+ - \hat{b}_{-j}) \tag{3.1a}$$

and

$$\hat{H}(z,t) = \sum_j \hat{h}_j = \sum_j -is_j\left(\frac{2\pi\omega_j}{V_o}\right)^{1/2} (\hat{b}_j^+ + \hat{b}_{-j}) \tag{3.1b}$$

where ω_j is the frequency of the j-th harmonic plane wave, and $s_j = \pm 1$ is the sign of j, positive for a forward-going (towards +z) and negative for a backward-going (towards -z) wave. The operators $\hat{b}_j^+(\hat{b}_j)$ create (annihilate) a photon in the j-th mode of the free-field, and follow Bose commutation relations. They differ however from the traditional photon creation (annihilation) operators, in that the spatial and temporal oscillations of the electromagnetic field are understood to be implicit in $\hat{b}_j^+(\hat{b}_j)$. The operators $\hat{e}_j$ and $\hat{h}_j$ defined in eqs. (3.1), are composed of a linear combination of creation and annihilation operators of opposite indices, and thus are not associated with a single wave, and do not represent physical observables of the field. They are simply a mathematical convenience, since they correspond to the individual components of the spatial Fourier expansion of the electric and magnetic fields respectively.

The free-space electric and magnetic field operators (3.1) permit us to convert all classical observables of the electromagnetic field into their quantum mechanical equivalent operators, both in free-space <u>and inside a material medium</u>. This conversion can be done by simply replacing the electric and magnetic fields in the classical definition of the observable by the corresponding operators of eqs. (3.1).

In particular, the energy density operator inside a linear medium is

$$\hat{u} = \frac{1}{8\pi}(\hat{E}^2 + \hat{H}^2 + 4\pi\chi\hat{E}^2) = \qquad (3.2a)$$

$$= \sum_j \hat{u}_j = \frac{1}{8\pi}\sum_j (\hat{e}_j\hat{e}_{-j} + \hat{h}_j\hat{h}_{-j} + 4\pi\chi\hat{e}_j\hat{e}_{-j}) \qquad (3.2b)$$

where we have assumed that the energy-density is distributed uniformly in space, and thus we eliminated from the sum, all terms that oscillate spatially. In terms of the free-field Bose creation and annihilation operators, each individual j-component of the energy-density operator may be written as,

$$\hat{u}_j = \frac{\omega_j}{2V_o}\{\hat{b}_j^+\hat{b}_j + \hat{b}_{-j}^+\hat{b}_{-j} - 2\pi\chi(\hat{b}_j^+ - \hat{b}_{-j})(\hat{b}_{-j}^+ - \hat{b}_j)\} \quad (3.3)$$

Thus, inside a linear dielectric the energy-density operator $\hat{u}$ is not diagonal when expressed in tems of free-field operators and includes a coupling between the two members of each pair of counter-propagating modes. This coupling consists of terms that do not conserve energy to first-order, and are of the form

$$2\pi\chi(\hat{b}_j^+\hat{b}_{-j}^+ + \hat{b}_j\hat{b}_{-j}) \tag{3.4}$$

These interaction terms may be eliminated, and the energy-density operator becomes diagonal when expressed in the refracted-wave basis set. The refracted-wave operators $\hat{B}_j^+/\hat{B}_j$ are related to the free-space operators $\hat{b}_j^+/\hat{b}_j$ through a Bogoliubov transformation. This transformation corresponds to the application of a unitary operator (similar to the "squeeze" operator used for example in the problem of parametric generation (Stoler, 1970; Stoler, 1971; Yuen, 1976) of the form

$$\hat{U} = \exp\{\sum_j \gamma\,(\hat{b}_j\hat{b}_{-j} - \hat{b}_j^+\hat{b}_{-j}^+)\} = \tag{3.5a}$$

$$= \exp\{\sum_j \gamma\,(\hat{B}_j\hat{B}_{-j} - \hat{B}_j^+\hat{B}_{-j}^+)\} \tag{3.5b}$$

where

$$\gamma = \frac{1}{4}\ln(1+4\pi\chi) = \frac{1}{2}\ln(n) \tag{3.6}$$

with n being the refractive index of the medium. This transformation relates the refracted-wave creation (annihilation) operators $\hat{B}_j^+(\hat{B}_j)$ to the corresponding free-field operators through

$$\hat{B}_j^+ = \hat{U}^{-1}\,\hat{b}_j^+\,\hat{U} = \cosh\gamma\,\hat{b}_j^+ - \sinh\gamma\,\hat{b}_{-j} \tag{3.7a}$$

$$\hat{B}_j = \hat{U}^{-1}\,\hat{b}_j\,\hat{U} = \cosh\gamma\,\hat{b}_j - \sinh\gamma\,\hat{b}_{-j}^+ \tag{3.7b}$$

$$\hat{b}_j^+ = \hat{U}\,\hat{B}_j^+\,\hat{U}^{-1} = \cosh\gamma\,\hat{B}_j^+ + \sinh\gamma\,\hat{B}_{-j} \tag{3.7c}$$

$$\hat{b}_j = \hat{U}\,\hat{B}_j\,\hat{U}^{-1} = \cosh\gamma\,\hat{B}_j + \sinh\gamma\,\hat{B}^{+}_{-j} \qquad (3.7d)$$

The relationship between the free-field and refracted-wave operators (eqs. 3.7) may also be expressed in terms of the refractive index, since

$$\cosh\gamma = \frac{n+1}{2\sqrt{n}} \qquad (3.8a)$$

$$\sinh\gamma = \frac{n-1}{2\sqrt{n}} \qquad (3.8b)$$

Inserting eqs.(3.7) into eq. (3.3), the energy density operator in the refracted-wave basis set may be obtained after some algebra as

$$\hat{u} = \sum_j \hat{u}_j = \sum_j \frac{n\,\omega_j}{V_o}\,\hat{B}^{+}_j\hat{B}_j \qquad (3.9)$$

The energy-density operator is thus diagonal in the refracted-wave basis set, and all its eigenvalues are related to the free-field energy-density eigenvalues by the refractive index n. This is exactly the same relationship as that obtained in classical optics (eq. 2.4), when solving the Maxwell equations for the field inside a refractive medium and in free-space. Thus, diagonalization of the energy-density operator in quantum optics is equivalent to the solution of the Maxwell equations in classical optics.

The proportionality of the energy-density eigenvalues to the refractive index implies that in order to have energy conservation upon propagation in quantum optics (or equivalently, in order to preserve the periodic-boundary conditions of the cavity of quantization), the volume of the refracted-wave quantization cavity, is related to the free-field quantization volume by

$$V = \frac{V_o}{n} \qquad (3.10)$$

as discussed in Section 2. With this quantization volume, the refracted-wave Hamiltonian is

$$\hat{\mathcal{H}} = V\ \hat{u} = \sum_j \omega_j \hat{B}_j^+ \hat{B}_j \tag{3.11}$$

and is identical in form and with the same eigenfrequencies as the free-space Hamiltonian.

The eqs. (3.7) relating the free-field and refracted-wave operators permit us to convert all operators to the refracted-wave basis set. Thus, substituting in eqs (3.1) the electric and magnetic field operators can be expressed in the refracted-wave basis set as

$$\hat{E}(z,t) = \sum_j -i\left(\frac{2\pi\omega_j}{nV_o}\right)^{1/2} (\hat{B}_j^+ - \hat{B}_{-j}) \tag{3.12a}$$

and

$$\hat{H}(z,t) = \sum_j -is_j\left(\frac{2n\pi\omega_j}{V_o}\right)^{1/2} (\hat{B}_j^+ + \hat{B}_{-j}) \tag{3.12b}$$

The refracted-wave momentum may be obtained in a similar fashion as

$$\hat{G} = V\ \hat{g} = \frac{1}{4\pi c}\frac{V_o}{n}(1+4\pi\chi)\sum_j \hat{e}_j \hat{h}_{-j} = \tag{3.13}$$

$$= \sum_j K_j \hat{B}_j^+ \hat{B}_j$$

where the j-th momentum eignenvalue

$$K_j = \sum_j s_j \frac{n\omega_j}{c} \tag{3.14}$$

is the wavevector of the j-th refracted wave.

Propagation of the electromagnetic field inside a linear medium, is completely described through the Hamiltonian (3.11) and momentum (3.13) operators, which permit the calculation of the temporal evolution and spatial progression of all operators and all states (classical and nonclassical) of the electromagnetic field. In particular, eq. (3.11) gives the time-evolution of any field operator, through the Heisenberg equation. For example, for the annihilation operator of the j-th mode we have,

$$\frac{\partial \hat{B}_j}{\partial t} = i\ [\hat{\mathcal{H}}, \hat{B}_j] = -i\,\omega_j \hat{B}_j \tag{3.15a}$$

Integrating this equation of motion, we obtain a time-harmonic evolution at frequency ω_j for the refracted waves,

$$\hat{B}_j(t) = e^{i\hat{\mathcal{H}}t}\ \hat{B}_j\ e^{-i\hat{\mathcal{H}}t} = \hat{B}_j(0)e^{-i\omega_j t} \tag{3.15b}$$

identical to that of the corresponding free-field modes. Similarly, the spatial progression of the field is obtained through eq. (2.2) which can be considered as a Heisenberg-like spatial equation of motion involing the momentum operator. Thus, for the j-th plane wave inside a linear medium,

$$\frac{\partial \hat{B}_j}{\partial z} = -i\ [\hat{G}, \hat{B}_j] = iK_j \hat{B}_j \tag{3.16a}$$

Integrating this equation we obtain, as in classical optics, the oscillatory spatial progression of a refracted wave,

$$\hat{B}_j(z) = e^{-i\hat{G}z}\ \hat{B}_j\ e^{i\hat{G}z} = \hat{B}_j e^{iK_j z} \tag{3.16b}$$

however with a spatial frequency (wavevector) renormalized by a factor of n with respect to the corresponding free-field wavevector, as expected from classical considerations.

Thus, as in classical optics, when a light beam traverses a linear dielectric, its frequency remains unchanged, while its wavevector is modified by the refractive index of the dielectric. The time-evolution of the refracted waves is thus independent of the medium of propagation: a time-harmonic wave remains time-harmonic at the same frequency. On the other hand, its spatial progression is altered, as its spatial frequency of oscillation (or equivalently its phase velocity) are changed by refraction.

4. REFLECTION IN PROPAGATIVE QUANTUM OPTICS

In classical optics, when a light beam is incident on the interface between empty space and a refractive medium (or between two refractive media) it splits into two beams: one is transmitted forward into the medium while the other is reflected back into empty space. The coexistence of the three waves, incident, transmitted and reflected, insures that the tangential components of the electric and magnetic fields are continuous across the interface.

Within the formalism of propagative quantum optics, the abrupt discontinuity in refractive index across the interface corresponds to an abrupt change of the energy density or momentum operators upon propagation, and thus may be treated by examining the spatial progression of the field and invoking the "sudden approximation" at the interface: A free-field wave incident on a refractive interface gets projected onto the corresponding waves of the refractive medium. Alternatively, the projection may be calculated by assuming that the continuity of the electric and magnetic fields across the interface (imposed by the Maxwell equations) corresponds to the continuity of the electric and magnetic field operators. Thus, the operators on either side of the interface are related through the unitary transformation (3.5) which relates, for example in eq. (3.7a), the forward-going the refracted-wave operators in the medium to the forward-going (incident) and backward-going (reflected) free-field operators. Writing eqs (3.7) in terms of the incident wave operators, and using eqs. (3.8), we obtain the operator equivalent of the Fresnel formulas of classical optics, that describe the reflection and refraction of light:

$$\hat{b}_j^+ = \frac{2\sqrt{n}}{n+1}\hat{B}_j^+ + \frac{n-1}{n+1}\hat{b}_{-j} \tag{4.1a}$$

$$\hat{b}_j = \frac{2\sqrt{n}}{n+1}\hat{B}_j + \frac{n-1}{n+1}\hat{b}_{-j}^+ \tag{4.1b}$$

$$\hat{B}_j^+ = \frac{2\sqrt{n}}{n+1}\hat{b}_j^+ - \frac{n-1}{n+1}\hat{B}_{-j} \tag{4.1c}$$

$$\hat{B}_j = \frac{2\sqrt{n}}{n+1}\hat{b}_j - \frac{n-1}{n+1}\hat{B}_{-j}^+ \tag{4.1d}$$

These equations describe quantum mechanically the process of reflection and refraction at an interface, and give quite naturally the familiar transmission and reflection coefficients as resulting directly from the Bogoliubov transformation that relates the free-field and refracted waves.

When a quasi-classical wavepacket of the j-th plane wave (Glauber, 1963)

$$|\alpha_I> = e^{-i\alpha(\hat{b}_j+\hat{b}_j^+)}|0> \tag{4.2}$$

is incident on the interface, it is projected according to eq. (4.1) and splits into a transmitted and a reflected quasi-classical wavepackets:

$$|\alpha_T,\alpha_R> = e^{-i\alpha\frac{2\sqrt{n}}{n+1}(\hat{B}_j+\hat{B}_j^+)}\; e^{-i\alpha\frac{n-1}{n+1}(\hat{b}_{-j}+\hat{b}_{-j}^+)}|0> \tag{4.3}$$

whose expectation values of the electric and magnetic fields

$$E_T = \frac{2}{n+1}E_I \qquad E_R = \frac{n-1}{n+1}E_I \tag{4.4a,b}$$

$$H_T = \frac{2n}{n+1}H_I \qquad H_R = -\frac{n-1}{n+1}H_I \tag{4.5a,b}$$

are identical to the corresponding classical expressions and thus satisfy the continuity condition for the tangential electric and magnetic fields.

The quantum mechanical Fresnel formulas (4.1) can also

describe the behavior of nonclassical light incident on a refractive interface. Thus, a single-photon state of the j-th free-field wave, which can be described as an oscillation of the electric field of the form

$$(\hat{b}_j + \hat{b}_j^+)\,|0\rangle \qquad (4.6)$$

will be transformed at a dielectric interface, according to eqs. (4.1), as

$$\frac{2\sqrt{n}}{n+1}\,(\hat{B}_j+\hat{B}_j^+)\,|0\rangle + \frac{n-1}{n+1}\,(\hat{b}_{-j}+\hat{b}_{-j}^+)\,|0\rangle \qquad (4.7)$$

to give an oscillation of a single-photon state either in the j-th refracted wave, or in the -j (reflected) free-field wave.

Thus, within this formalism, the quantum mechanical Fresnel formulas (4.1) permit to describe beamsplitting in a manner completely analogous to that of classical optics, while at the same time they account for the quantum statistical properties of both the reflected and the refracted waves.

5. CONCLUSIONS

In quantum optics a general state of the electromagnetic field consists of a function of the electric and magnetic field operators applied onto the electromagnetic vacuum state. For states that have a classical analogue, the electric and magnetic field operators present an expectation value that correponds to the complex oscillating classical field. The propagation of such states through a polarizable medium, may thus be described through the macroscopic Maxwell equations which can be used to calculate both the time evolution and the spatial progression of these expectation values. For nonclassical light on the other hand, the expectation values of the electric and magnetic fields do not characterize completely the state of the field. Thus, in order to describe the propagation of such states, it is necessary to calculate both the temporal evolution and the spatial progression of the field operators.

Propagative quantum optics is an extension of the conventional theory of quantum optics that relies on the momentum operator of the electromagnetic field, for the calculation of the spatial progression of light, in addition to the Hamiltonian that describes the temporal evolution. In this way, this theory treats the spatial and temporal coordinates of the field on the same footing. The medium is introduced quasi-phenomenologically through its induced polarization. A key feature of propagative quantum optics is that the field is quantized in a cavity whose physical dimensions depend on the medium it contains, in such a way that periodic conditions are obeyed at its boundaries irrespective of its material contents. The preservation of periodic-boundary conditions in all media, insures that the electromagnetic energy is conserved in the field as light propagates through different media, as is the case also in classical optics. At the same time, this theory incorporates a quantum mechanical version of the Maxwell equations: diagonalization of the energy density operator is equivalent to the solution of the Maxwell equations. Thus, propagative quantum optics is the direct quantum mechanical equivalent of the conventional (classical) propagative optics and reduces to it in the appropriate limit.

When applied to a linear medium, this theory gives the results expected by analogy with classical optics. The induced polarization of the medium does not affect the temporal evolution of the field, but changes its spatial progression: it gives rise both to a refracted wave in the medium (with renormalized wavevector) and to a reflected wave at the interface. The transmission and reflection coefficients, familiar from classical optics, are given within this formalism directly by the transformation that diagonalizes the energy density operator. The direct correspondance of the results of this theory with the well-known results of classical (linear) optics demonstrates the validity of this approach, and permits us to apply this formalism to the description of the evolution of the quantum statistics of nonclassical states of light upon their propagation in linear or nonlinear media.

REFERENCES

Abram, I. 1987, Phys. Rev. A 35, 4661

Abram, I. 1988, Phys. Rev. A, in press

Glauber, R.J. 1963, Phys. Rev. 131, 2766

Peierls, R. 1985, Highlights of Condensed-Matter Physics, Ed. F. Bassani, F. Fumi and M.P. Tosi, North-Holland, Amsterdam, pp. 237-255

Stoler, D. 1970, Phys. Rev. D 1, 3217

Stoler, D. 1971, Phys. Rev. D 4, 1925

Yuen, H.P. 1976, Phys. Rev. A 13, 2226

Yee, T.K and Gustafson, T.K. 1978, Phys. Rev. A 18, 1597

MODELS FOR PHASE-INSENSITIVE QUANTUM AMPLIFIERS

G L MANDER, R LOUDON AND T J SHEPHERD

1. INTRODUCTION

Study of the properties of linear optical amplifiers is motivated by both academic and technological interest in the field. In pure science, experimental tests of theories must incorporate measuring apparatus, the last essential quantum-mechanical stage of which is a high-gain amplifier to boost the signal into one which may be regarded as classical for measurement purposes. In applied science, amplifier theory is of relevance to the communications industry where, for example, semiconductor laser amplifiers may have use as receiver pre-amplifiers and non-regenerative repeaters in optical fibre links.

The existence of squeezed light, and its potential use to carry optical information with an improved signal-to-noise ratio relative to coherent light, has led to a need to know the degree of gain that is possible in the system before the squeezing, and also other quantum properties of light such as antibunching, are lost. A determining factor in these calculations is the amount of necessarily-added quantum noise inherent to the system. Although there exist phase-sensitive linear amplifiers that add no noise to one quadrature phase of the field, for example, degenerate parametric amplifiers, in general the signal-to-noise ratio will be degraded by the amplification process, and quantum statistical properties may be washed out by the randomizing effect of the noise field.

We concentrate our attention here on reviewing and extending the work on the particular class of linear amplifiers known as inverted population amplifiers. We discuss first the fundamental requirements demanded by quantum mechanics from any theory of linear amplifiers, and then describe the basic model for the atomic amplifier in a closed cavity. We consider the extension of this model to allow explicit incoherent coupling of the internal cavity field to external fields, and treat finally an open system with coherent field coupling. We investigate in particular the conditions under which the amplified field retains quantum properties such as squeezing.

2. FUNDAMENTAL REQUIREMENTS OF LINEAR AMPLIFIER THEORY

Since no system can be truly isolated from the rest of the universe, being surrounded at the very least by a vacuum field, there will always be fluctuations and dissipation in the system due to its interaction with the containing reservoir of its environment. Lax (1966) suggested that the reservoir could be eliminated from the calculations, provided that the mean equations of motion incorporate the reservoir-induced frequency shifts and dissipation, and also provided that a suitable effective operator noise source is added. The effect of this noise source is clearly irremovable, and forces an amplifier to add a certain minimum amount of noise to any signal it processes. Naively, and ideally, one might wish to write, for an output field described by the annihilation operator $\hat{a}_{out}$ and derived from an input field described by $\hat{a}_{in}$;

$$\hat{a}_{out} = u\hat{a}_{in} \tag{2.1}$$

so that the mean photon number amplification is

$$\langle n_{out} \rangle = |u|^2 \langle n_{in} \rangle \tag{2.2}$$

and the power gain (loss) is $G = |u|^2$. However, from the discussion above, and also from the unitarity requirements of quantum mechanics, we must write,

$$\hat{a}_{out} = u\hat{a}_{in} + \hat{F} \tag{2.3}$$

where $\hat{F}$ is the noise operator, which can be expressed in terms of the internal amplifier modes. The input and output fields then obey the necessary commutation relations, viz

$$[\hat{a}_{out}, \hat{a}^{\dagger}_{out}] = [\hat{a}_{in}, \hat{a}^{\dagger}_{in}] = 1 \tag{2.4}$$

if the noise operator satisfies

$$[\hat{F}, \hat{F}^{\dagger}] = 1 - G \tag{2.5}$$

Now expectation values of hermitean squares of operators must be real and non-negative, so that

$$\langle \hat{F}\hat{F}^{\dagger} \rangle \geq 0$$
$$\langle \hat{F}^{\dagger}\hat{F} \rangle \geq 0 \tag{2.6}$$

and (5) and (6) together give the fundamental theorem for phase-insensitive linear amplifiers in Caves' (1982) form,

$$\langle \hat{F}^{\dagger}\hat{F} \rangle \geq G - 1 \tag{2.7}$$

This is the mathematical expression of the fact that such an amplifying device must necessarily degrade the signal: the added noise only vanishes for G = 1, i.e. a passive component.

If the boson fields carry coherent information then the expectation values $\langle \hat{a}_{in} \rangle$, $\langle \hat{a}_{out} \rangle$ are non-vanishing. Defining their fluctuations as the symmetrical correlation functions, for any boson operator $\hat{b}$

$$\langle |\Delta\hat{b}|^2 \rangle = \tfrac{1}{2}\langle \hat{b}^{\dagger}\hat{b} + \hat{b}\hat{b}^{\dagger} \rangle - \langle \hat{b} \rangle \langle \hat{b}^{\dagger} \rangle \tag{2.8}$$

which gives $\langle |\Delta\hat{b}|^2 \rangle \geq \tfrac{1}{2}$. From the basic amplifier equation (3) then

$$\langle |\Delta\hat{a}_{out}|^2 \rangle = G\langle |\Delta\hat{a}_{in}|^2 \rangle + \langle |\Delta\hat{F}|^2 \rangle \tag{2.9}$$

Using the equivalent noise factor of Caves (1982),

$$A = \frac{\langle |\Delta\hat{F}|^2 \rangle}{G} \geq \tfrac{1}{2}\left(1 - \frac{1}{G}\right) \tag{2.10}$$

we have,

$$\langle |\Delta\hat{a}_{out}|^2 \rangle \geq G(\tfrac{1}{2} + \tfrac{1}{2}) \tag{2.11}$$

This shows explicitly that, for symmetrically ordered fluctuation operators, the amplification process adds noise equivalent to one half photon at the amplifier input, where the minimum input fluctuation is also one half photon. If the fluctuation operators are normally ordered, then the equivalent noise at the amplifier input is one photon.

Any viable theory of linear amplifiers must satisfy the fundamental theorem. Usually one works with the quantum dynamics of the amplifying degrees of freedom, so that use of the effective quantum noise sources can be avoided. This also has the advantage that the amplifier parameters may be related directly to the physical mechanism of the process. For example, in inverted population amplifiers, the gain is directly expressible in terms of the atomic inversion. The amplifier added noise may also be expressed in terms of the atomic properties. It is precisely this noise that breaks the time symmetry inherent in the basic equations of quantum mechanics. That amplification should be intrinsically irreversible can be seen by considering any particular atom in the collection comprising the amplifier. When the atom decays, the remaining atoms form a photon sink. If a photon coherent with the field is emitted and then absorbed by another atom, it may be reradiated spontaneously, with its subsequent loss from the beam. Spontaneous emission being random, this process is clearly not reversible: indeed any attempt to run the amplifier backwards (Glauber, 1986), that is, to attenuate the beam, would only result in the further degradation of the signal.

3. THE CLOSED SYSTEM

This model was proposed in 1963 by Gordon, Walker and Louisell (1963), and has since been greatly used in investigation of the effect of optical amplification on the properties of quantum light fields. We will therefore look in some detail at its main features.

3.1 Model

The system is composed of a single radiation mode in a lossless cavity coupled to a large number of nearly independent atoms (Fig. 1). Each atom is assumed to have a finite or infinite set of equally spaced energy levels, and it is supposed that there are sufficiently many atoms with transition frequencies close to the field frequency that the atom density of states may be taken as continuous. Only atoms with transition frequencies near the field frequency couple with any strength to the field. Further, the atoms are taken to be weakly coupled to a heat bath, so that they have a Boltzmann distribution. The statistical distribution of the atoms is taken, by virtue of their large number and weak coupling to the field, to be constant over the interaction time. These assumptions serve to ensure linearity, i.e. non-saturation of the amplifier.

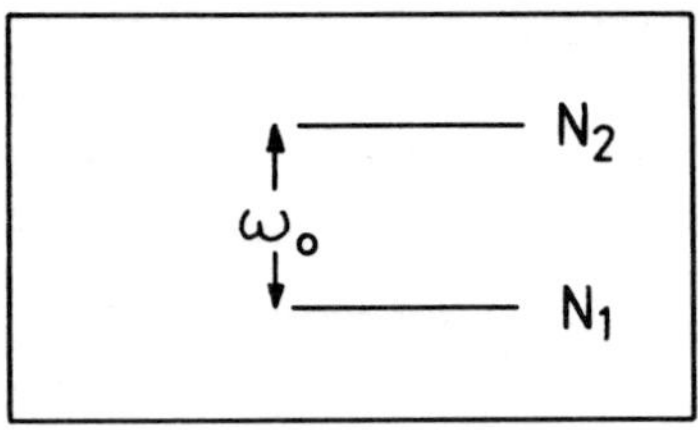

Fig. 1

Closed Cavity

N 2-level atoms : $N_1 + N_2 = N$

3.2 Analysis

If $\hat{a}(t)$ is the annihilation operator of the internal field, and $\sigma_j^-(t)$ is the lowering operator for the jth atom then the Hamiltonian of the system has the usual form in the rotating-wave approximation,

$$\hat{H} = \hbar\omega_o \hat{a}^+\hat{a} + \tfrac{1}{2}\sum_{j=1}^{N} \hbar\omega_j \sigma_j^z + \sum_{j=1}^{N} \hbar\kappa_j (\hat{\sigma}_j^+\hat{a} + \hat{\sigma}_j^-\hat{a}^+) \tag{3.1}$$

where $\hat{\sigma}_j^z$ is the atomic inversion operator, and κ_j is the atom-field coupling constant. The relevant commutation relations give the Heisenberg equations of motion for the field and atomic operators as

$$\frac{d\hat{a}}{dt} = -i\omega_o a - i\sum_j \kappa_j \hat{\sigma}_j^-$$

$$\frac{d\hat{\sigma}_j^-}{dt} = -i\omega_j \hat{\sigma}_j^- + i\kappa_j \langle\hat{\sigma}_j^z\rangle \hat{a} \tag{3.2}$$

where it is assumed that the population inversion is externally maintained in order to obtain linearized equations. We are interested in the evolution of the cavity mode, and the

Heisenberg equation may be solved in the Wigner-Weisskopf approximation to give the result

$$\hat{a}(t) = u(t)\hat{a}(0) + \sum_{j=1}^{N} v_j(t)\hat{\sigma}_j^- \tag{3.3}$$

with

$$u(t) = \exp[-i\omega_o t + (N_2-N_1)\gamma_L t] \tag{3.4}$$

$$\gamma_L = \frac{\pi}{N}\kappa^2(\omega_o)\rho(\omega_o)$$

$\rho(\omega_o)$ = atomic density of states.

The quantity $v_j(t)$ varies linearly with the coupling constant, and has the property

$$\sum_j |v_j(t)|^2 \langle\hat{\sigma}_j^+\sigma_j^-\rangle = P(G-1) \tag{3.5}$$

where now the gain G is defined by

$$G = |u(t)|^2 = \exp[2(N_2-N_1)\gamma_L t] \tag{3.6}$$

and P is the population factor

$$P = \frac{N_2}{(N_2-N_1)} \tag{3.7}$$

The result obtained is time-varying, so one chooses to identify $\hat{a}(t = 0)$ as the input mode and $\hat{a}(t)$ as the output mode of the system,

$$\hat{a}_{in} = \hat{a}(0)$$

$$\hat{a}_{out} = \hat{a}(t) \tag{3.8}$$

The input-output relation (3.3) then takes the form of the standard result (2.3) for phase-insensitive amplifiers. Notice that the system will only show gain under the physically satisfactory condition that $N_2 > N_1$.

Unitarity is obeyed only on the average,

$$\langle[\hat{a}_{out},\hat{a}_{out}^\dagger]\rangle_{atoms} = 1 \tag{3.9}$$

consistent with the linearizing approximation.

The properties of the output field may be evaluated in terms of those of the input using the following relation,

$$\langle[\hat{a}^{\dagger}_{out}]^{r}[\hat{a}_{out}]^{r+s}\rangle$$

$$= \sum_{q=0}^{r} \binom{r}{q}\binom{r+s}{q+s}(r-q)!\,\bar{n}_{ch}^{\,r-q}\langle[u^{*}\hat{a}^{\dagger}_{in}]^{q}[u\hat{a}_{in}]^{q+s}\rangle \qquad (3.10)$$

where $\bar{n}_{ch} = P(G-1)$. This equation has a well-known form which indicates that the noise operator is associated with a field having thermal statistics and a mean number of photons $\bar{n}_{ch}$.

The interesting question to ask now is whether we can amplify a field using this system and reduce the shot-noise fluctuation of the field. Performing a direct detection experiment, we call the input signal

$$S_{in} = \langle\hat{a}^{\dagger}_{in}\hat{a}_{in}\rangle = \bar{n}_{in}, \qquad (3.11)$$

the output signal

$$S_{out} = G\bar{n}_{in} = \bar{n}_{out} - \bar{n}_{ch} \qquad (3.12)$$

and the noise on the input and output

$$\text{Noise} = \text{var } n = n^{(2)} + \bar{n} - \bar{n}^{2}$$

Note that this noise is not the same as the amplitude noise in (2.8) to (2.11). If we evaluate the enhancement of the signal-to-noise ratio of the ouput relative to the input in terms of a parameter R,

$$R = \frac{SNR_{out}}{SNR_{in}} \quad : \quad SNR = \frac{S^{2}}{\text{Noise}} \qquad (3.13)$$

then we find

$$R = \left[1 + \left(1 - \frac{1}{G}\right)\left(\frac{\bar{n}_{in}(2P-1) + P\left[\frac{1}{G} + P\left(1 - \frac{1}{G}\right)\right]}{\text{var } n_{in}}\right)\right]^{-1} \qquad (3.14)$$

This quantity is always less than or equal to unity for the allowed range of parameters for amplification so that enhancement is not possible, and the output photon distribution is always broadened relative to the input, i.e. the signal is always degraded.

3.3 Preservation of Non-Classical Effects

We consider first the possibility of preserving sub-poissonian statistics after amplification. A field is sub-poissonian if it has a second-order coherence less than one. We find that the output field will be sub-poissonian if

$$G < \left[1 + \frac{\bar{n}_{in}}{P}\left(1 - \sqrt{2-g_{in}^{(2)}}\right)\right]^{-1} \qquad (3.15)$$

where $g_{in}^{(2)}$ is the second-order coherence of the input field. Under optimum operating conditions, that $N_1 = 0$ $(P = 1)$, the maximum gain one may have is two.

Similarly for squeezing, if one defines an operator

$$\hat{X}(\chi) = \hat{a}^{+}e^{i\chi} + \hat{a}\,e^{-i\chi} \qquad (3.16)$$

and evaluate the quantity

$$\begin{aligned} Q_{out} &= \tfrac{1}{2}\langle:(\Delta\hat{X}(\chi))^2:\rangle \\ &= GQ_{in} + P(G - 1) \end{aligned} \qquad (3.17)$$

then the field will be squeezed if $Q_{out} < 0$. The minimum value of P is 1 and of Q_{in} is $-\frac{1}{2}$ so that again the maximum gain which retains the quantum property after amplification is two. These results may be compared with those of Hong et al (1985) who show that, for a single-mode linear amplifier, in general the maximum intensity gain which retains quantum characteristics in the output field is two.

4. THE OPEN SYSTEM WITH INCOHERENT FIELD COUPLING

A major disadvantage of the Louisell model is its non-stationary nature. Since there are no losses in the system the gain is unbounded and becomes infinite for long enough times. This is clearly unrealistic, and the introduction of loss mechanisms through coupling the cavity mode to external fields is a necessary modification of the model. In the remainder of this paper we discuss two ways of achieving stationarity, beginning with a population approach suggested by Shepherd and Jakeman (1987).

4.1 Model

We again take a cavity containing a collection of two-level atoms and supporting a single mode of radiation, but now the cavity is open to admit entry and exit of external photon fields (fig. 2).

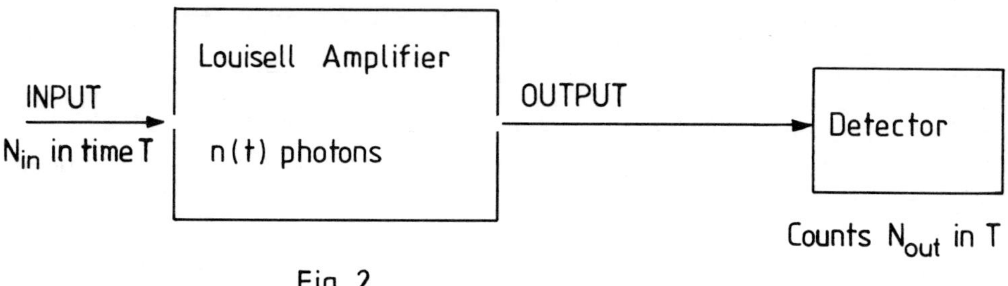

Fig. 2

Incoherently Coupled Open System

4.2 Analysis

Following Friberg and Mandel (1983) one can derive the forward Kolmogorov equation for the discrete Markov process taking place in the cavity. The stimulated emission and absorption are represented by population-dependent (branching) birth and death processes, with rates λ and μ respectively, and the spontaneous emission by a population-independent immigration process at rate λ. The external fields are then regarded as further immigration and death processes with respect to the internal field, at rates η and ν. The result is

$$\frac{dP(N,n;t)}{dt} = [\lambda N+\nu]P(N-1,n;t) + \mu(N+1)P(N+1,n;t) \qquad (4.1)$$

$$- [\lambda(N+1)+\mu N+\nu+\eta N]P(N,n;t) + \eta(N+1)P(N+1,n-1;t)$$

where $P(N,n;t)$ is the joint probability distribution for N photons in the cavity <u>and</u> n photoelectrons counted in the interval (O,t). It is assumed that the input photons just add to the cavity population, and that the statistics of the input are determined by the time at which the individual photons are added, i.e. that the coupling of the internal and external fields is incoherent. The equation is most conveniently manipulated using a distribution-generating function defined as

$$Q(s,s';t) = \sum_{N=0}^{\infty} \sum_{n=0}^{\infty} P(N,n;t)(1-s)^N(1-s')^n \qquad (4.2)$$

The partial differential equation governing the evolution of $Q(s,s';t)$ may be solved analytically (Shepherd and Jakeman, 1987) and the solution, taken with the Markov properties of the system, enables the evaluation of relevant statistical monents and correlation functions. We will not go into the derivations in detail, but will concentrate on the main results.

The gain of the system is found to be

$$G = \frac{\eta}{\mu + \eta - \lambda} \qquad (4.3)$$

so that amplification is possible provided $\lambda > \mu$. Considering the enhancement or otherwise of the signal-to-noise ratio after processing, the parameter R defined in (3.13) is here found to be

$$R = \frac{G}{(1 + \lambda/\nu)[1 + 2PG(G-1)f(\gamma)]} \qquad (4.4)$$

where $f(\gamma)$ is a monotonically increasing function of γ,

$$f(\gamma) = \frac{\exp(-2\gamma) + 2\gamma - 1}{2\gamma} \qquad (4.5)$$

and

$$\gamma = (\mu + \eta - \lambda)T/2 \qquad (4.6)$$

T is the detector integration time. From this Shepherd and Jakeman find that there is indeed enhancement in the output SNR relative to the input, but only if the integration time for the statistics is less than the photon number correlation time of the amplifier. This is because the amplifier provides temporal smoothing, exhibiting a delay in its response to a signal due to the finite time needed to re-excite an atom after it has decayed. The reconciliation of intensity gain with steady-state behaviour achieved here is a result of the additional damping on the internal mode provided by the photodetection process. The main drawback of this discrete population approach is the lack of phase information available, meaning, amongst other things, that one cannot examine squeezing effects with this model.

5. THE OPEN SYSTEM WITH COHERENT FIELD COUPLING

In our final model the cavity is again open, but with the external fields multimode and coherently coupled to the internal mode using the coupling scheme of Collett and Gardiner (1984).

5.1 Model

The quantum Langevin equation for the system is taken as the starting point for the derivation of input/output relations. We consider a double-ended cavity, supporting a single mode of oscillation described by the annihilation operator $\hat{a}(t)$, and interacting with multimode external fields (fig. 3).

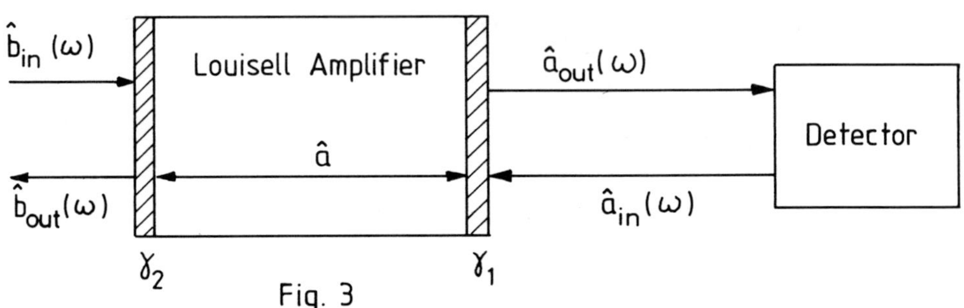

Fig. 3

Coherently Coupled Open System

5.2 Analysis

The external fields have commutation relations

$$[\hat{b}_{in}(\omega),\hat{b}^{\dagger}_{in}(\omega')] = [\hat{a}_{out}(\omega),\hat{a}^{\dagger}_{out}(\omega')] = \delta(\omega-\omega'). \qquad (5.1)$$

If γ_1 and γ_2 measure the end-mirror transmissivities, i.e. represent the cavity damping, then the quantum Langevin equation is

$$\frac{d\hat{a}}{dt} = \frac{-i}{\hbar}[\hat{a},\hat{H}_{sys}] - \tfrac{1}{2}(\gamma_1+\gamma_2)\hat{a}(t) + \gamma_1^{\frac{1}{2}}\hat{a}_{in}(t) + \gamma_2^{\frac{1}{2}}\hat{b}_{in}(t) \qquad (5.2)$$

with the usual Hamiltonian (3.1),

$$\hat{H}_{sys} = \hbar\omega_o\hat{a}^{+}\hat{a} + \tfrac{1}{2}\sum_j\hbar\omega_j\hat{\sigma}_j^{+} + \sum_j\hbar\kappa_j(\hat{\sigma}_j^{+}\hat{a} + \hat{\sigma}_j^{-}\hat{a}^{+}) \qquad (5.3)$$

in the rotating wave approximation. We wish to solve the Langevin equation for the internal mode, and so eliminate the atomic operators via the appropriate Heisenberg equation, making similar linearizing assumptions to those of Louisell. We can use the boundary conditions at the right hand mirror,

$$\gamma_1^{\frac{1}{2}}\,\hat{a}(t) = \hat{a}_{in}(t) + \hat{a}_{out}(t) \tag{5.4}$$

to find the result for the output field operator (Mander and Loudon, to be published)

$$\hat{a}_{out}(t) = -\hat{a}_{in}(t) + \int_{-\infty}^{t} d\tau\, e^{(-i\omega_o-\Gamma)(t-\tau)}\,[\gamma_1\hat{a}_{in}(\tau)+(\gamma_1\gamma_2)^{\frac{1}{2}}\hat{b}_{in}(\tau)] + \gamma_1^{\frac{1}{2}}\sum_j \frac{k_j\hat{\sigma}_j^{-}\,e^{-i\omega_j t}}{\omega_j-\omega_o+i\Gamma} \tag{5.5}$$

with $\Gamma = \frac{1}{2}(\gamma_1+\gamma_2) - (N_2-N_1)\gamma_L$

as the inverse cavity lifetime, and γ_L given by (14). Reading from left to right, we have the output field expressed in terms of reflected input field, amplified input fields and the noise (spontaneous emission) field.

As with the first model, unitarity is obeyed on the average,

$$\langle[\hat{a}_{out}(t),\hat{a}^{\dagger}_{out}(t')]\rangle_{atoms} = \delta(t-t'). \tag{5.6}$$

5.3 Flux

We can now investigate the properties of the output field in terms of those of the input, modified by the amplification process. For continuous external fields it is most convenient to deal in terms of flux, so that we first define a flux operator

$$\hat{f}(t) = \hat{a}^{\dagger}(t)\hat{a}(t) = \frac{1}{2\pi}\int d\omega \int d\omega'\; e^{i(\omega-\omega')t}\hat{a}^{\dagger}(\omega)\hat{a}(\omega') \tag{5.7}$$

then the mean flux f(t) is given by

$$f(t) \equiv \langle\hat{f}(t)\rangle \equiv \int d\omega f(\omega) \tag{5.8}$$

where $f(\omega)$ is the flux per unit frequency bandwidth at frequency ω.

The mean output flux is then

$$f_{out}(\omega) = \langle \hat{a}^\dagger_{out}(t)\hat{a}_{out}(t)\rangle = G(\omega)f_{in}(\omega) + f_{ch}(\omega) \tag{5.9}$$

where the gain

$$G(\omega) = \frac{\gamma_1\gamma_2}{(\omega-\omega_o)^2 + \Gamma^2}$$

and the chaotic flux

$$f_{ch}(\omega) = \frac{N_2\gamma_1\gamma_L/\pi}{(\omega-\omega_o)^2 + \Gamma^2} \tag{5.10}$$

and here and in what follows we assume the right hand input to be a vacuum field for simplicity.

5.4 Output spectral density

Following Caves (1982) we define the output spectral density by

$$S^O(\omega)\delta(\omega-\omega') = \tfrac{1}{2}\langle \hat{a}^\dagger_{out}(\omega)\hat{a}_{out}(\omega') + \hat{a}_{out}(\omega')\hat{a}^\dagger_{out}(\omega)\rangle \tag{5.11}$$

and find from this that we may write

$$S^O(\omega) = G(\omega)[S^I(\omega) + S^A(\omega)] \tag{5.12}$$

where

$$S^I(\omega) = 2\pi\ f_{in}(\omega) + \tfrac{1}{2}$$

$$S^A(\omega) = \frac{2ne\gamma_L}{\gamma_2} + \tfrac{1}{2}\left|\frac{1}{G(\omega)} - 1\right| \tag{5.13}$$

The fundamental theorem for a multimode, phase-insensitive system requires

$$S^A(\omega) \geq \tfrac{1}{2}\left|1 - \frac{1}{G(\omega)}\right| \tag{5.14}$$

Compare (2.10). There will be amplification, i.e. gain greater than unity, if

$$2\pi f_{ch}(\omega) \geq G(\omega) - 1 \tag{5.15}$$

as required.

5.5 Preservation of non-classical effects

We look finally at the ability of this model to retain quantum fields after amplification. Defining an operator $\hat{X}(\chi,t)$ such that

$$\hat{X}(\chi,t) = \tfrac{1}{2}(e^{i\chi - i\omega_o t}\hat{a}^\dagger(t) + e^{-i\chi + i\omega_o t}\hat{a}(t)) \tag{5.16}$$

which is a generalisation of (3.16), then the light is squeezed if, for some phase angle ,

$$<(\Delta\hat{X}(\chi,t))^2> \;\equiv\; \int_{-\infty}^{\infty} d\tau \{X(\chi,\tau)X(\chi,0)> - <\hat{X}(\chi,\tau)><\hat{X}(\chi,0)>\} \;<\tfrac{1}{4} \tag{5.17}$$

It is possible to show that the output field is squeezed if

$$G < \frac{\gamma_1\gamma_2}{[\gamma_2<(\Delta\hat{X}(\chi))^2>_{in} + \tfrac{1}{4}\gamma_2 + \tfrac{1}{2}\gamma_1 + N_1\gamma_L]^2} \tag{5.18}$$

Under the optimum conditions that $N_1 = 0$ and

$$\frac{\gamma_1}{\gamma_2} = 2<(\hat{X}(\chi))^2>_{in} + \tfrac{1}{2} \tag{5.19}$$

then

$$G_{max} = \frac{1}{2<(\Delta\hat{X}(\chi))^2>_{in} + \tfrac{1}{2}} \tag{5.20}$$

Since the minimum value of $<(\Delta\hat{X}(\chi))^2>_{in} = 0$, then again the maximum gain that preserves squeezing after amplification is two.

6. CONCLUSIONS

We have looked at three models of laser amplifiers operating below threshold. In particular we have examined the ability of the systems to enhance the signal-to-noise ratio of an injected signal, and found that in general any signal is degraded by the amplification process. One exception occurs in the case of an open system with incoherently coupled internal and external fields, when enhancement will be seen if the detection integration time for the photon statistics is less than the amplifier coherence time. Since squeezed light could play an important role in future communication systems, in which amplifiers to boost the signal will be an essential part, we investigated the limitations of the systems in their ability to preserve quantum features of the input field after processing. We found, as predicted by Hong et al. (1985) that such phase-insensitive atomic amplifiers destroy all quantum statistical properties of the field if the gain exceeds the low value of two. Clearly these amplifiers will be of limited use in any practical system where the gain needed at the repeater is typically of the order of 1000.

ACKNOWLEDGEMENTS

G.L. Mander thanks the Science and Engineering Research Council for financial support in the form of a research studentship.

REFERENCES

Caves, C.M. 1982, Phys. Rev. D26 1817
Collett, M.J. and Gardiner, C.W. 1984, Phys. Rev. A30 3 1386
Friberg, S. and Mandel, L. 1983, Opt. Comm. 46 141
Glauber, R.J. 1986, Frontiers in Quantum Optics, Ed. E.R. Pike and S. Sarkar, Hilgar, Bristol, pp. 534-582
Gordon, J.P., Walker, L.R. and Louisell, W.H. 1963, Phys. Rev. 130 2 806
Hong, C.K., Friberg, S. and Mandel, L. 1985, J. Opt. Soc. Am. B2 494
Lax, M. 1966, Phys. Rev. 145 1 110
Shepherd, T.J. and Jakeman, E. 1987, J. Opt. Soc. Am. B4 1860

QUASIPROBABILITIES BASED ON SQUEEZED STATES

F HAAKE AND M WILKENS

We generalize quasiprobabilities based on coherent states to ones based on squeezed states. Especially, the representation of an operator as a diagonal mixture of squeezed states defines an analogue of the Glauber-Sudarshan P-function. The new quasiprobabilities can be useful tools in describing processes the intrinsic quantum nature of which forbids the existence of the familiar P-function. We illustrate our concepts for a linearized model of subharmonic generation.

1. INTRODUCTION

Many experiments in quantum optics involve a linear or nonlinear oscillator subjected to driving by classical fields as well as the dissipative influence of heat baths. The theoretical tractability of such a system often hinges on a choice of a represention for the density operator particularly well suited to the dynamics under study.

Various quasiprobabilities based on coherent states $|\alpha\rangle$ have proven successful such representations, among them the socalled Q function (Glauber, 1963; 1966)

$$Q(\alpha) = \frac{1}{\pi} \langle \alpha|\rho|\alpha \rangle \ . \tag{1.1}$$

A whole one parameter family of similar quasiprobabilities, $W_\epsilon(\alpha)$, is defined so as to give $Q(\alpha)$ upon convolution with a Gaussian of width ϵ (Graham et al., 1968; Cahill and Glauber, 1969a; 1969b),

$$Q(\alpha) = \frac{1}{\pi\epsilon} \int d^2\beta \; e^{-|\alpha-\beta|^2/\epsilon} \; W_\epsilon(\beta) \, , \tag{1.2}$$

$$0 \leq \epsilon \leq 1 \, .$$

The potentially most singular member of this family, the Glauber-Sudarshan P function, pertains to $\epsilon = 1$ while Q itself obviously resides at the other extreme, $\epsilon = 0$. Halfway in between P and Q at $\epsilon = {}^1/_2$ lies the Wigner function (Wigner, 1932). Each member of the family allows the calculation of means of normally ordered products of creation and annihilation operators,

$$\langle \, a^{+m} \, a^n \, \rangle = \int d^2\alpha \left[\alpha^* + (1-\epsilon) \frac{\partial}{\partial\alpha}\right]^m \left[\alpha + (1-\epsilon) \frac{\partial}{\partial\alpha^*}\right]^n W_\epsilon(\alpha) \, . \tag{1.3}$$

We shall here concern ourselves with generalizing (1,2,3) by using squeezed instead of coherent states as a basis. Such a generalization seems worthwhile in view of numerous recent and ongoing experiments in which the quantum fluctuations in an oscillator are anisotropically squeezed with respect to those in a coherently excited oscillator. For a review see Walls (1983). One may hope that the dynamics of a "squeezed" oscillator takes a simpler and more easily tractable form if described by squeezed-state based quasiprobabilities rather than by the coherent-state based $W_\epsilon(\alpha)$.

2. SQUEEZED VS COHERENT STATES

The following brief review of coherent and squeezed states will mainly serve to set up our notation. We shall employ the well known unitary displacement and squeezing operators defined such that (Cahill and Glauber, 1969a,b; Stoler, 1970; Yuen, 1976)

$$\begin{aligned} D(\alpha)\ a\ D^{+}(\alpha) &= a - \alpha \\ S(\eta)\ a\ S^{+}(\eta) &= a\ \mathrm{ch}\ r + a^{+}\ e^{i2\theta}\ \mathrm{sh}\ r \\ &\equiv a(\eta) \end{aligned} \tag{2.1}$$

where both the displacement α and the squeezing parameter η are complex numbers, the latter to be represented most often by polar coordinates

$$\eta = r\ e^{i2\theta}\ . \tag{2.2}$$

With the help of these unitary operators the coherent state $|\alpha\ \rangle$ and the squeezed state $|\alpha;\eta\ \rangle$ are generated from the vacuum state $|0\ \rangle$ as

$$|\alpha\ \rangle = D(\alpha)\ |0\ \rangle\ , \quad |\alpha;\eta\ \rangle = D(\alpha)\ S(\eta)\ |0\ \rangle\ . \tag{2.3}$$

Evidently, the squeezed state $|\alpha;\eta\ \rangle$ reduces to the coherent state $|\alpha\ \rangle$ for $\eta = 0$. Moreover, by using a consequence of (2.1),

$$\begin{aligned} S^{+}(\eta)\ D(\alpha)\ S(\eta) &= D(\alpha(\eta)) \\ \alpha(\eta) &= \alpha\ \mathrm{ch}\ r + \alpha^{*}\ e^{i2\theta}\ \mathrm{sh}\ r\ , \end{aligned} \tag{2.4}$$

we can rewrite the squeezed state as

$$|\alpha;\eta\ \rangle = S(\eta)\ D(\alpha(\eta))\ |0\ \rangle \tag{2.5}$$

and deduce

$$a(\eta)\ |\alpha;\eta\rangle = \alpha(\eta)\ |\alpha;\eta\rangle\ , \tag{2.6}$$

i.e. that the squeezed state is a coherent state with respect to the squeezed annihilator $a(\eta)$ defined in (2.1).

The physical meaning of η becomes obvious from the squeezed-state variances of the observable

$$X_\Theta = e^{-i\Theta}\ a + e^{i\Theta}\ a^+ \tag{2.7}$$

and the correspondingly defined $X_{\Theta + \pi/2}$,

$$\begin{aligned} \langle\ \alpha;\eta|(\Delta\ X_\Theta)^2|\alpha;\eta\ \rangle &= e^{-2r} \\ \langle\ \alpha;\eta|\left[\Delta\ X_{\Theta+\pi/2}\right]^2|\alpha;\eta\ \rangle &= e^{+2r}\ . \end{aligned} \tag{2.8}$$

The phases Θ and $\Theta + \pi/2$ thus defines the directions in the complex α plance with respect to which the squeezed state $|\alpha;\eta\rangle$ describes maximum squeezing and maximum stretching, respectively, of quantum fluctuations. The modulus r, on the other hand, determines the degree of that squeezing and stretching.

The representation (2.5) of the squeezed state immediately yields the generalization of (i) the scalar product of different coherent states to that of the corresponding squeezed states,

$$\langle\ \beta;\eta|\alpha;\eta\ \rangle = \exp\left\{\beta^*(\eta)\ \alpha(\eta) - \frac{1}{2}\ |\beta(\eta)|^2 - \frac{1}{2}\ |\alpha(\eta)|^2\right\}\ , \tag{2.9}$$

and (ii) the resolution of unity in terms of coherent states to the one in terms of squeezed states,

$$1 = \int \frac{d^2\alpha}{\pi} |\alpha;\eta\rangle\langle\alpha;\eta| \; . \qquad (2.10)$$

3. SQUEEZED-STATE BASED QUASIPROBABILITIES

The obvious generalization of the coherent-state based Q-function (1.1) is

$$Q(\alpha(\eta);\eta) = \frac{1}{\pi} \langle\alpha;\eta|\rho|\alpha;\eta\rangle \; . \qquad (3.1)$$

Its Fourier transform $\chi(\xi;\eta)$, defined by

$$Q(\alpha(\eta);\eta) = \frac{1}{\pi^2} \int d^2\xi \; e^{-\xi\alpha^*(\eta)+\xi^*\alpha(\eta)} \chi(\xi;\eta) \; , \qquad (3.2)$$

allows to represent the density operator ρ in a form normally ordered with respect to the squeezed operators $a(\eta)$ and $a^+(\eta)$ as

$$\rho = \int \frac{d^2\xi}{\pi} e^{-\xi a^+(\eta)} e^{\xi a(\eta)} \chi(\xi;\eta) \; . \qquad (3.3)$$

Conversely, $\chi(\xi;\eta)$ can be directly obtained from the density operator through

$$\chi(\xi;\eta) = \mathrm{tr}\; e^{-\xi^* a(\eta)} e^{\xi a^+(\eta)} \rho \; , \qquad (3.4)$$

as is easily verified by inserting the resolution of unity (2.10) in (3.4). Needless to say, all of the above reduces to well known lore about the coherent-state based Q function $Q(\alpha) = Q(\alpha(0);0)$ for $\eta = 0$.

It will be important to express the coherent-state based Q function in terms of its squeezed-state based generalization.

To achieve that goal we start from (3.4). By rearranging the antinormal order with respect to $a(\eta)$ and $a^+(\eta)$ into antinormal order with respect to a and a^+ with the help of the Baker Campbell Hausdorf formula we find a relation between the Fourier transforms of $Q(\alpha)$ and $Q(\alpha(\eta);\eta)$,

$$\chi(\xi) = e^{-\frac{1}{2}\xi\xi^* + \frac{1}{2}\xi(\eta)\,\xi^*(\eta)}\,\chi(\xi(\eta);\eta)\ . \tag{3.5}$$

Upon taking Fourier transforms we arrive at

$$Q(\alpha) = \int \frac{d^2\xi}{\pi}\,\frac{d^2\beta}{\pi}\,e^{-\xi\alpha^* + \xi^*\alpha + \xi(\eta)\beta^* - \xi^*(\eta)\beta} \cdot e^{-\frac{1}{2}\xi\xi^* + \frac{1}{2}\xi(\eta)\xi^*(\eta)} \cdot Q(\beta;\eta)\ . \tag{3.6}$$

Note that we cannot reverse the order of the integrations in (3.6) since the Gaussian in (3.5) and (3.6) is unbounded in the complex ξ plane.

We now proceed to generalize the whole family $W_\epsilon(\alpha)$ of coherent-state based quasiprobabilities to a family $W_\epsilon(\alpha(\eta);\eta)$ of squeezed-state based ones by

$$Q(\alpha(\eta);\eta) = \frac{1}{\pi\epsilon}\int d^2\beta\ e^{-|\beta - \alpha(\eta)|^2/\epsilon}\ W_\epsilon(\beta;\eta)\ ,$$

$$0 \le \epsilon \le 1\ .$$

It is easy to see, by using the overlap formula (2.9), that the special case $\epsilon = 1$ defines the squeezed-state based P function which allows to represent the density operator as a mixture of squeezed states

$$\rho = \int d^2\alpha\ P(\alpha(\eta);\eta)\ |\alpha;\eta\rangle\langle\alpha;\eta|\ . \tag{3.8}$$

As is the case for the family $W_\epsilon(\alpha)$ we continuously interpolate between the Q function and the P function when let the width parameter ϵ grow from 0 to 1. Incidentally, $\epsilon = {}^1/_2$ again corresponds to a Wigner function.

By combining (3.6), (3.7), and (1.2) we easily establish an important relation between $W_\epsilon(\alpha)$ and $W_\epsilon(\alpha(\eta);\eta)$,

$$\begin{aligned} W_\epsilon(\alpha) = \int \frac{d^2\xi}{\pi}\frac{d^2\beta}{\pi}\, & e^{-\xi(\alpha^*-\beta^*) + \xi^*(\alpha-\beta)} \\ & \cdot e^{-(\frac{1}{2}-\epsilon)\,\xi\xi^* + (\frac{1}{2}-\epsilon')\,\xi(\eta)\,\xi^*(\eta)} \\ & \cdot W_{\epsilon'}(\beta(\eta);\eta)\ , \end{aligned} \tag{3.9}$$

$$0 \leq \epsilon,\epsilon' \leq 1\ .$$

The Gaussian in (3.9), in contrast to that in (3.6) can be made bounded by restricting ϵ and ϵ' as

$$\begin{aligned} & (\epsilon' - \tfrac{1}{2})\, e^{-2r} \geq (\epsilon - \tfrac{1}{2}) \quad \text{for} \quad \epsilon' \geq {}^1/_2 \\ \text{or} \quad & (\tfrac{1}{2} - \epsilon')\, e^{2r} \leq (\tfrac{1}{2} - \epsilon) \quad \text{for} \quad \epsilon' \leq {}^1/_2\ . \end{aligned} \tag{3.10}$$

With that condition met the ξ integral in (3.9) may be carried out and yields $W_\epsilon(\alpha)$ as a convolution of $W_{\epsilon'}(\alpha(\eta);\eta)$ with a certain Gaussian $G(\alpha;\epsilon,\epsilon',\eta)$ the width matrix of which depends on the parameters ϵ,ϵ', and η.

4. QUASIPROBABILITIES REPRESENTING A PURE SQUEEZED STATE

If the density operator is a projector on a squeezed state,

$$\rho = |\alpha_0;\eta\rangle\langle\alpha_0;\eta| \ , \tag{4.1}$$

we can immediately specify the P function

$$\begin{aligned} P(\alpha(\eta);\eta) &= \delta^2(\alpha(\eta) - \alpha_0(\eta)) \\ &= \delta^2(\alpha - \alpha_0) \ . \end{aligned} \tag{4.2}$$

From (3.7) we then get all the other squeezed-state based $W_\epsilon(\alpha(\eta);\eta)$ as the Gaussians

$$W_\epsilon(\alpha(\eta);\eta) = \frac{1}{\pi(1-\epsilon)} \exp\left\{- |\alpha(\eta) - \alpha_0(\eta)|^2/(1-\epsilon)\right\} , \tag{4.3}$$

$$0 \leq \epsilon \leq 1 \ .$$

It is quite important to realize, by using (3.9) that not all the $W_\epsilon(\alpha)$ pertaining to (4.1) exist. In fact, the existence requirement is (3.10) with $\epsilon = 1$ and $\epsilon' \to \epsilon$,

$$\epsilon \leq \frac{1}{2}(1 + e^{-2r}) \ . \tag{4.4}$$

The $W_\epsilon(\alpha)$ with ϵ obeying (4.4) are all Gaussian in form. Obviously, the condition (4.4) excludes the coherent-state based P function; the pure squeezed state (4.1) can not be represented as a mixture of coherent states.

5. SUBTHRESHOLD SUBHARMONIC GENERATION

We would now like to illustrate the above squeezed-state techniques for a simple dynamical model of subharmonic generation. For the sake of simplicity we confine ourselves to the subthreshold regime where the oscillation to be down converted can be treated parametrically. The corresponding well known master equation for the density operator of the down converted oscillation reads (Drummond et al., 1979; Milburn and Walls, 1981)

$$\dot{\rho} = -i\,[H,\rho] + \Lambda\,\rho \tag{5.1}$$

with the Hamiltonian

$$H = \delta\, a^{+}a + \frac{i}{2}\, f\,(a^{2} - a^{+2}) \tag{5.2}$$

and the damping generator Λ

$$\Lambda\,\rho = \gamma \left\{ [a,\rho\, a^{+}] + [a\,\rho, a^{+}] \right\} . \tag{5.3}$$

The three parameters δ, f, and γ are a detuning between the free oscillation and the external field, the amplitude of the external field, and the damping constant.

By using the techniques described in Sec. 3 we can translate the master equation (5.1) into a c number partial differential for the squeezed-state based P function $P(\alpha;\eta)$ with η a free parameter,

$$\begin{aligned} \dot{P}(\alpha,t) &= \mathcal{L}\, P(\alpha,t) \\ \mathcal{L} &= \frac{\partial}{\partial\alpha}\left\{(i\Omega + \gamma)\,\alpha + F\,\alpha^{*}\right\} \\ &\quad + \frac{\partial^{2}}{\partial\alpha^{2}}\left\{ \gamma\, e^{i2\theta}\, \mathrm{ch}\, r\, \mathrm{sh}\, r - \frac{1}{2} F\right\} \\ &\quad + \frac{\partial}{\partial\alpha\, \partial\alpha^{*}}\, \gamma\, \mathrm{sh}^{2} r \\ &\quad + c.c. \;. \end{aligned} \tag{5.4}$$

$$\Omega = \delta \cosh 2r + f \sinh 2r \sin 2\theta$$

$$F = -\,i\,\delta \sinh 2r\, e^{i2\theta} + f\left[\cosh^{2} r - \sinh^{2} r\, e^{i4\theta}\right] .$$

This is a Fokker Planck equation of a Gaussian stochastic process provided the diffusion matrix formed by the coefficients of the second-order derivatives is nonnegative.

It is quite important to realize that the diffusion matrix of (5.4) has one positive and one negative eigenvalue if we set $\eta = 0$, i.e. try to work with the coherent-state based P function. A negative diffusion coefficient amounts to a shrinking of the width of $P(\alpha)$ along the direction defined by the corresponding eigenvector and thus to squeezing.

By requiring the diffusion matrix to be positive we confine the squeezing parameter η to a finite singly connected region in the complex η plane. We may try to use the remaining freedom in choosing η to give as simple a structure as possible to the Fokker Planck equation (5.4). A choice of special physical significance results from requiring the stationary solution of (5.4) to be isotropic in the α plane, i.e. to be a function of the product $\alpha\alpha^*$ only. The stationary squeezing in our process is then fully accounted for by the set of basis states $|\alpha;\eta\rangle$ with η fixed to

$$
\begin{aligned}
e^{i2\theta} &= (\gamma - i\delta)/\sqrt{\gamma^2 + \delta^2} \\
\mathrm{sh}^2 r &= \frac{\sqrt{\gamma^2 + \delta^2} - \sqrt{\gamma^2 + \delta^2 - f^2}}{2\sqrt{\gamma^2 + \delta^2 - f^2}} \, .
\end{aligned}
\tag{5.5}
$$

The isotropic stationary density reads

$$
P(\alpha;\eta) = (\pi\, \mathrm{sh}^2 r)^{-1} \exp\left\{ -\alpha\alpha^*/\mathrm{sh}^2 r \right\} \tag{5.6}
$$

and yields the mean values

$$\langle a \rangle = \langle a^{+} \rangle = 0$$

$$\langle a^2 \rangle = -\frac{f}{2(\gamma + i\delta)}(1 + 2\langle a^{+}a \rangle) \qquad (5.7)$$

$$\langle a^{+}a \rangle = \frac{f^2}{2(\gamma^2 + \delta^2 - f^2)} .$$

We should note that the model considered here looses physical meaning for a driving force f above the threshold value $f_{thr} = \sqrt{\gamma^2 + \delta^2}$ as is manifest in the results (5.5-7). The model nonetheless allows a nice illustration of the usefulness of squeezed-state based quasiprobabilities.

We gratefully acknowledge support by the Sonderforschungsbereich 237 der Deutschen Forschungsgemeinschaft.

REFERENCES

Cahill, K. E. and Glauber, R. J., 1969a, Phys. Rev. 177 1857
Cahill, K. E. and Glauber, R. J., 1969b, Phys. Rev. 177 1882
Drummond, P. D., McNeal, K. J. and Walls, D. F., 1979, Optics Comm. 28 255
Glauber, R. J., 1963, Phys. Rev. 130 2529
Glauber, R. J., 1966, Phys. Rev. 131 2766
Graham, R., Haake, F., Haken, H. and Weidlich, W., 1968, Z. Phys. 213 21
Milburn, G. and Walls, D. F., 1981, Optics Comm. 39 401
Stoler, D., 1970, Phys. Rev. D1 3217
Walls, D. F., 1983, Nature 306 141
Wigner, E. P., 1932, Phys. Rev. 40 749
Yuen, H. P., 1976, Phys. Rev. A13 2226

INDEX